Young Children's Developing Understanding of the Biological World

This book explores current research on young children's beliefs and knowledge about the biological world – otherwise known as 'folkbiology'. Contributors discuss factors that shape the development of folkbiological knowledge, as well as possible interventions designed to counteract cognitive biases that can interfere with the development of scientifically informed reasoning about natural phenomena.

Taken together, the papers provide insights into the contributions of cognitive biases to the development of biological misunderstandings and into the life experiences and contexts that can contribute to or impede accurate learning of biological concepts. As part of a wider literature, the insights provided by the authors are relevant to the design of educational experiences that will foster children's exploration and further their understanding of life science ideas.

The chapters in this book were originally published as a special issue of *Early Education and Development*.

Peter J. Marshall is Professor of Psychology and Chair of the Department of Psychology at Temple University, USA. His research interests center on cognitive and social development in infancy and early childhood. He has published on various areas of developmental science including the ways in which young children understand the functioning of the brain. His theoretical writings also include a biologically inspired approach to the interface of psychology and neuroscience, with an emphasis on the embodied nature of mental life.

Kimberly Brenneman is a program officer for education at the Heising-Simons Foundation, USA. Prior to joining the foundation in 2015, she was research faculty at Rutgers University's National Institute for Early Education Research. At Rutgers, she led projects focused on curricular and instructional practices and resources that foster science, technology, engineering, and mathematics learning for young children in school and home settings. As an educational consultant, she contributes to the development of educational media resources to support preschool-aged math and science learners.

Young Children's Developing Understanding of the Biological World

Edited by
Peter J. Marshall and Kimberly Brenneman

Routledge
Taylor & Francis Group
LONDON AND NEW YORK

First published 2018
by Routledge
2 Park Square, Milton Park, Abingdon, Oxon, OX14 4RN, UK

and by Routledge
711 Third Avenue, New York, NY 10017, USA

Routledge is an imprint of the Taylor & Francis Group, an informa business

British Library Cataloguing in Publication Data
A catalogue record for this book is available from the British Library

ISBN13: 978-1-138-56463-3

Typeset in MinionPro
by diacriTech, Chennai

Publisher's Note
The publisher accepts responsibility for any inconsistencies that may have arisen during the conversion of this book from journal articles to book chapters, namely the possible inclusion of journal terminology.

Disclaimer
Every effort has been made to contact copyright holders for their permission to reprint material in this book. The publishers would be grateful to hear from any copyright holder who is not here acknowledged and will undertake to rectify any errors or omissions in future editions of this book.

Contents

Citation Information

The chapters in this book were originally published in *Early Education and Development*, volume 27, issue 8 (November–December 2016). When citing this material, please use the original page numbering for each article, as follows:

Introduction

Young Children's Developing Understanding of the Biological World
Peter J. Marshall and Kimberly Brenneman
Early Education and Development, volume 27, issue 8 (November–December 2016) pp. 1103–1108

Chapter 1

"Inhabitants of the Earth": Reasoning About Folkbiological Concepts in Wichi Children and Adults
Andrea S. Taverna, Douglas L. Medin, and Sandra R. Waxman
Early Education and Development, volume 27, issue 8 (November–December 2016) pp. 1109–1129

Chapter 2

Generation Conservation: Children's Developing Folkbiological and Moral Conceptions of Protecting Endangered Species
Jolina H. Ruckert
Early Education and Development, volume 27, issue 8 (November–December 2016) pp. 1130–1144

Chapter 3

Direct Experience With Nature and the Development of Biological Knowledge
Sarah E. Longbottom and Virginia Slaughter
Early Education and Development, volume 27, issue 8 (November–December 2016) pp. 1145–1158

Chapter 4

Young Chinese Children's Justifications of Plants as Living Things
Ying Tao
Early Education and Development, volume 27, issue 8 (November–December 2016) pp. 1159–1174

Chapter 5

Children's Cognitive and Behavioral Reactions to an Autonomous Versus Controlled Social Robot Dog
Nadia Chernyak and Heather E. Gary
Early Education and Development, volume 27, issue 8 (November–December 2016) pp. 1175–1189

Chapter 6

Are Prompts Provided by Electronic Books as Effective for Teaching Preschoolers a Biological Concept as Those Provided by Adults?
Gabrielle A. Strouse and Patricia A. Ganea
Early Education and Development, volume 27, issue 8 (November–December 2016) pp. 1190–1204

Chapter 7

Changing Minds With the Story of Adaptation: Strategies for Teaching Young Children About Natural Selection
Natalie Emmons, Hayley Smith, and Deborah Kelemen
Early Education and Development, volume 27, issue 8 (November–December 2016) pp. 1205–1221

Chapter 8

Children's Ability to Learn Evolutionary Explanations for Biological Adaptation
Andrew Shtulman, Cara Neal, and Gabrielle Lindquist
Early Education and Development, volume 27, issue 8 (November–December 2016) pp. 1222–1236

Chapter 9

Using Animals to Teach Children Biology: Exploring the Use of Biological Explanations in Children's Anthropomorphic Storybooks
Megan Geerdts, Gretchen Van De Walle, and Vanessa LoBue
Early Education and Development, volume 27, issue 8 (November–December 2016) pp. 1237–1249

Chapter 10

Body and soul: Do children distinguish between foods when generalizing biological and psychological properties?
Jean-Pierre Thibaut, Simone P. Nguyen, and Gregory L. Murphy
Early Education and Development, volume 27, issue 8 (November–December 2016) pp. 1250–1262

Notes on Contributors

Kimberly Brenneman is a program officer for education at the Heising-Simons Foundation, USA.

Nadia Chernyak is a Postdoctoral Fellow at the Department of Psychology, Boston College, USA.

Natalie Emmons is a Postdoctoral Researcher at the Department of Psychological and Brain Sciences, Boston University, USA.

Patricia A. Ganea is Associate Professor at the Department of Applied Psychology and Human Development, University of Toronto, Canada.

Heather E. Gary graduated from the Department of Psychology, University of Washington, USA.

Megan Geerdts is Assistant Professor at the Department of Psychology, William Paterson University, USA.

Deborah Kelemen is Professor at the Department of Psychological and Brain Sciences, Boston University, USA.

Gabrielle Lindquist completed her Bachelor's Degree in Psychology at the Department of Psychology, Occidental College, USA.

Vanessa LoBue is Associate Professor at the Department of Psychology, Rutgers University, USA.

Sarah E. Longbottom is a PhD student at the Early Cognitive Development Centre, University of Queensland, Australia.

Peter J. Marshall is Professor of Psychology and Chair of the Department of Psychology at Temple University, USA.

Douglas L. Medin is Professor at the Department of Psychology, Northwestern University, USA.

Gregory L. Murphy is Professor of Psychology at the Department of Psychology, New York University, USA.

Cara Neal completed her Bachelor's Degree in Psychology at the Department of Psychology, Occidental College, USA.

Simone P. Nguyen is Professor at the Department of Psychology, University of North Carolina at Wilmington, USA.

Jolina H. Ruckert is Visiting Assistant Professor at the Department of Psychology, Lewis & Clark College, USA.

Andrew Shtulman is Associate Professor of Psychology and Cognitive Science at the Department of Psychology, Occidental College, USA.

Virginia Slaughter is Principal Researcher at the Early Cognitive Development Centre, University of Queensland, Australia.

Hayley Smith is based at the Psychology Department, University of Massachusetts Boston, USA.

Gabrielle A. Strouse is Assistant Professor at the Division of Counseling and Psychology in Education, University of South Dakota, USA.

Ying Tao is Associate Professor at the School of Education Science, Nanjing Normal University, China.

Andrea S. Taverna is a Researcher Member of the National Research Council, Argentina.

Jean-Pierre Thibaut is Professor of Developmental Psychology at the University of Bourgogne Franche-Comté, France.

Gretchen Van De Walle is Associate Professor at the Department of Psychology, Rutgers University, USA.

Sandra R. Waxman is a Professor at the Department of Psychology, Northwestern University, USA.

INTRODUCTION

Young Children's Developing Understanding of the Biological World

Peter J. Marshall and Kimberly Brenneman

The purpose of this special issue of *Early Education & Development* is to showcase research on young children's understanding of the living world. In this Introduction our aim is to tie together the various articles in the issue by identifying key themes that run through multiple articles. We also provide some reflections on implications of the studies for policy and practice, including prescriptions for future work on the integrative study of children's biological understanding in the context of the early education classroom.

One important theme in the study of children's biological understanding is *folkbiology*, which refers to the theoretical frameworks of beliefs and expectations about living things that emerge in the absence of formal instruction (Solomon & Zaitchik, 2012). There is increasing interest in the factors that shape the development of folkbiological knowledge (Gelman, 2009; Medin, Waxman, Woodring, & Washinawatok, 2010), a theme that is apparent throughout the special issue. The most prominent example is the article by Taverna, Medin, and Waxman (2016), who provide new insights from their work with the Wichi, an indigenous group from a remote forest region in Argentina. This study documents how Wichi children organize the natural world into categories based on taxonomic or perceptual similarities, on ecological factors such as shared habitat, and on beliefs about animacy and spiritual status. The emergence of these categories guides a view of the natural world that is distinctly less anthropocentric (i.e., less human centered) than the view adopted by children in many Western societies, especially urban ones.

The theme of anthropocentrism is also apparent in the article by Ruckert (2016), who examines the intersection of folkbiological thinking with the development of moral reasoning about environmental concerns. Working in the Pacific Northwest of the United States, Ruckert examines children's responses to questions about values, obligations, and rights concerning endangered species. She contrasts anthropocentric thinking with a biocentric form of moral reasoning in which the natural world has moral standing independent of its value to humans. Ruckert finds that between 7 and 10 years of age, children's thinking shifts from human-centered concerns to more biocentric reasoning.

Continuing the theme of influences on children's thinking about the natural world, the review by Longbottom and Slaughter (2016) examines how type of upbringing (urban vs. rural) and pet ownership affect children's reasoning about the relation of humans to other animals, the connections between different species, and basic knowledge about animal biology. After showing how differences in children's biological understandings are often associated with their direct experiences with outdoor experiences and with animals, the authors suggest that intervention research is needed on underlying mechanisms. Related anecdotal evidence that outdoor experiences influence children's biological understanding comes from the article by Tao (2016). In a Chinese sample of children ages 4 to 6 years, children who consistently applied a biological justification for plants as living things often mentioned their hands-on experience and involvement with growing flowers and vegetables with their parents or grandparents.

In addressing the question of whether children see plants as alive, Tao (2016) touches on a fundamental question related to biological understanding: What distinguishes between living organisms and nonliving objects? Preschool-age children already demonstrate particular ways of thinking about the distinction between living and nonliving things (Inagaki & Hatano, 2006). A related area of research applies similar questions to robots, which are not alive but which can seemingly act in an autonomous fashion. In this issue, Chernyak and Gary (2016) examine how young children interacted with a robot dog. In one condition, children were given indications that the robot was controlled by an experimenter, whereas children in the other condition were given the impression that the robot was autonomous. Children were more likely to ascribe emotional states, desires, and sentience to the robot that was presented as being autonomous. It is interesting, given Longbottom and Slaughter's (2016) review, that children who had pets were more likely to treat the robot dog in a prosocial manner.

The findings of Chernyak and Gary (2016) add to research examining how young children interact with robots (e.g., Tanaka, Cicourel, & Movellan, 2007) by suggesting that the way in which robots are presented can affect children's interactions with them, with implications for supporting different kinds of learning experiences. This work is part of a growing literature that explores how children learn from interactions with electronic devices, from smartphones and tablets to humanoid robots. One question in this area is how adult mediation of these interactions, such as verbal scaffolding and structuring of the experience with the device, affects children's learning. Strouse and Ganea (2016) explore this by studying children's understanding of camouflage as a biological concept after they interacted with an electronic book (e-book) under different scaffolding conditions. Their results suggest that children learned from their interactions with the e-book but that for children with lower initial levels of language skill, prompting by a live adult was particularly useful as a motivation for children to test and revise their biological understandings.

Another theme in the special issue concerns the ways in which the folkbiological understandings that children bring to the early learning context can complicate the learning of life science concepts. Two articles address this issue in the context of how children can come to understand the concept of adaptation that lies at the heart of the theory of evolution. Emmons, Smith, and Kelemen (2016) explain that the difficulty of explaining adaptation is accentuated by particular cognitive biases that characterize the way in which young children come to view the world of living things. One such bias concerns children's susceptibility to teleological reasoning, or the tendency to assume that "natural phenomena exist to perform functions and that the natural world is agentive and operates in intentional, designed, and purpose-driven ways" (Emmons et al., 2016, p. 1206). Because this bias may underlie many of the misunderstandings of evolution that are prevalent in adults, it is important to understand how it arises in children and what factors might accentuate or diminish it.

Emmons et al. (2016) build on their prior work by engaging a diverse sample of children in their testing of a storybook intervention about adaptation that is aimed at encouraging a less teleological view of natural phenomena. They report that when presented with age-appropriate explanations, young children could engage in theoretical learning about the process of adaptation. In a related study in the current issue, Shtulman, Neal, and Lindquist (2016) broadly replicate the main findings of Kelemen et al. (2014). However, the youngest children in the study of Shtulman and colleagues showed only minimal learning, perhaps due to the fact that the level of prompting was less extensive than in the prior work.

The studies of Emmons et al. (2016) and Shtulman et al. (2016) demonstrate the value of interventions using storybooks that have been designed to counteract the cognitive biases that interfere with the development of scientifically informed reasoning about natural phenomena. However, many of the storybooks that are available for families with young children may actually play into or reinforce these biases. In the current issue, Geerdts, Van De Walle, and LoBue (2016) analyze the types of causal explanations that commercial storybooks about animals provide about two biological concepts: inheritance and the transmission of illness. Their conclusion is that storybooks almost uniformly focus on social-emotional experiences rather than presenting scientifically accurate accounts. According to the authors, this focus encourages anthropomorphism, or the tendency to attribute human characteristics to nonhuman species, which in turn interferes with accurate biological reasoning and encourages an anthropocentric view of the world.

As well as being relevant to abstract concepts such as evolution, the development of children's thinking about biology is also relevant to children's learning about everyday matters such as health and nutrition (Siegal & Petersen, 1999). Thibaut, Nguyen, and Murphy (2016) explore the latter issue, showing how research in cognitive development can intersect with the study of how children view the biological properties of objects. Thibaut et al. use an inductive inference task to examine children's reasoning about whether taxonomically related foods (e.g., strawberries and watermelon) as opposed to script-related foods (e.g., strawberries and whipped cream) will cause similar effects when eaten. As noted by the authors, the findings have applied significance for understanding children's food choices and eating behavior.

The development of children's understanding of biological processes has also been studied in terms of contamination and the transmission of infectious diseases, the understanding of which requires reasoning about unobservable entities such as germs or toxins (Legare, Wellman, & Gelman, 2009). One question here is the extent to which the environment provides inputs about causal mechanisms for infection or contagion. As noted previously, children's storybooks do not provide much information about mechanisms for illness transmission (Geerdts et al., 2016). Neither, it seems, do the adults in young children's lives. Toyama (2016)[1] examined how preschool teachers talk to young children about contamination, specifically about cleanliness and hygiene at mealtimes. Although teachers put great emphasis on hygiene (e.g., washing hands before eating), they did not give precise explanations as to why hygienic practices were important. Despite this, children's own explanations of why breaking hygiene norms would lead to sickness were surprisingly sophisticated given the minimal explanations that they heard from teachers.

The work of Toyama (2016) suggests that young children are sometimes capable of generating accurate explanations of unobservable biological phenomena despite limited guidance from adults. As other articles in the special issue indicate, particular experiences (e.g., time outdoors, pet ownership) may contribute to the accuracy of the understandings that children build. Anthropocentrism, in contrast, seems to arise when deep experiences with the natural world are less available and when anthropomorphic representations of animals and nature are more available. Misunderstandings of biological principles may well emerge from internal cognitive biases, but available media representations and informal adult input are unlikely to provide accurate information to offset these biases.

The articles in the current issue provide insights into the development of these misunderstandings and into the experiences that seem likely to contribute to or impede accurate learning of biological concepts. As part of a wider literature, these articles are relevant to the development of educational experiences in life science for young children, a topic that is the focus of the concluding section of the current article.

Implications for policy and practice

Taverna et al. (2016) situate their work in an specific policy context in Argentina, arguing that, despite good intentions, curricular decisions there are often implemented without an understanding of cultural influences on scientific reasoning and understanding. They argue that teaching materials reflecting cultural knowledge are required if educational policies—and the associated curricula they aim to set into place—are to be successful. Although other authors in this issue do

[1]The article by Toyama (2016) was originally meant to be part of this special issue but was inadvertently published in an earlier issue. This article can be found at http://tandfonline.com/doi/full/10.1080/10409289.2015.1036347.

not call out education policy directly, many of the articles share the conclusion that experience shapes biological understandings and that the folkbiological frameworks that children bring to early learning settings must be understood and acknowledged as both resources and potential barriers to learning science.

As noted previously, one framework that children may bring to school is a human-centered approach to reasoning about the natural world. However, work presented here also shows that children who live in rural environments or in indigenous cultures do not seem as likely to privilege human beings as the inductive base when reasoning about the natural world. Similarly, experience with pets and with gardening is related to more sophisticated biological reasoning. A general awareness that direct experience with living things helps to build biological understanding likely underlies efforts to include class pets, gardening experiences, and animals such as chicks and butterflies in early education classrooms. However, although the presence of plants and animals in classrooms is a useful step toward fostering the development of biological understanding, it is not clear whether these opportunities are leveraged by teachers in order to build deeper, more comprehensive understandings. Instead these might be isolated activities, with children not being meaningfully involved in the care and observation of animals and plants over time.

Scientifically inaccurate messages from media—paired with sparse opportunities to engage directly with nature—are strong candidates for influencing the understandings that children build about animals and plants and their relationship to humans. In turn, these understandings influence the reasoning that children apply when thinking about related topics, such as conservation and animal rights. From an educational practice standpoint, new emphases in the Common Core standards for English language arts (National Governors Association Center for Best Practices & Council of Chief State School Officers, 2010) on the use of informational texts in U.S. classrooms, and efforts to incorporate similar kinds of texts into preschool curricula (e.g., Sarama, Brenneman, Clements, Duke, & Hemmeter, 2015), provide potential avenues for offsetting the anthropomorphic messages that children are commonly exposed to.

In terms of what such texts should look like, Geerdts et al. (2016) argue that engaging stories with factual information and accurate biological causal mechanisms can and should be offered to young children. How much easier would it be to foster accurate biological understandings—and head off inaccurate ones—if the books that are ubiquitous in many children's lives presented accurate descriptions of concepts such as contagion, inheritance, and evolution? Shtulman et al. (2016) and Emmons et al. (2016) present evidence that suggests the potential of such books to foster emerging understandings of evolutionary principles, an accurate and coherent conception of which eludes many adults. Collaborative efforts between authors and publishers of children's books and developmental researchers to create strong, engaging stories that include accurate causal information could fill a need by providing informal educational resources that blend learning and entertainment and that harness the power of narrative to foster scientific understanding.

The role of adults in children's emerging understandings of biology is touched on in various articles in the current issue. Parents, caregivers, and teachers play critical roles in guiding children's scientific learning, but many adults carry their own folkbiological understandings that influence how they present information to children, the language they use to do so, and the way they answer children's questions. In attention to informing curricular materials for children's learning, the studies in the special issue might also inform educational resources targeting the adults who serve as children's partners in learning. Emmons et al. (2016) found that of the few parents who were willing to respond to a question about evolutionary principles, very few were able to generate an accurate answer. Ideally, incorporating explanations for adults into educational materials for children or providing parent or teacher guides is important so that these most influential adult sources of children's information become resources for accurate biological information. In this way, children get consistent information from educational media and from their parents, caregivers, and teachers, a conceptual redundancy that can foster biology learning.

Although all of the authors in this special issue discuss educational applications for their work, real progress on the creation of curricular materials for informal and formal education settings is most likely to be made when practitioners and educational researchers are partners with developmentalists in the creation of research agendas. That is, end users—educators—should be involved as collaborators in the formulation of research questions so that the data generated answer the pressing questions that teachers, administrators, and education policymakers have about how children learn scientific principles and how best to structure learning environments to support this learning. When these stakeholders are included as collaborators, research findings are more likely to be applicable for practice and policy and more likely to find their way to the people who can best use them to support children's learning through the development of research-informed standards, curricular resources, and learning opportunities.

Acknowledgments

We thank Sandra Twardosz for her encouragement and support during the preparation of this special issue. The opinions stated in this article are our own and are not necessarily reflective of those of the Heising-Simons Foundation.

References

Chernyak, N., & Gary, H. E. (2016). Children's cognitive and behavioral reactions to an autonomous versus controlled social robot dog. *Early Education & Development*, *27*, 1175–1189. doi:10.1080/10409289.2016.1158611

Emmons, N., Smith, H., & Kelemen, D. (2016). Changing minds with the story of adaptation: Strategies for teaching young children about natural selection. *Early Education & Development*, *27*, 1205–1221. doi:10.1080/10409289.2016.1169823

Geerdts, M., Van De Walle, G., & LoBue, V. (2016). Using animals to teach children biology: Exploring the use of biological explanations in children's anthropomorphic storybooks. *Early Education & Development*, *27*, 1237–1249. doi:10.1080/10409289.2016.1174052

Gelman, S. A. (2009). Learning from others: Children's construction of concepts. *Annual Review of Psychology*, *60*, 115–140. doi:10.1146/annurev.psych.59.103006.093659

Inagaki, K., & Hatano, G. (2006). Young children's conception of the biological world. *Current Directions in Psychological Science*, *15*, 177–181. doi:10.1111/cdir.2006.15.issue-4

Kelemen, D., Emmons, N. A., Schillaci, R. S., & Ganea, P. A. (2014). Young children can be taught basic natural selection using a picture-storybook intervention. *Psychological Science*, *25*, 893–902. doi:10.1177/0956797613516009

Legare, C. H., Wellman, H. M., & Gelman, S. A. (2009). Evidence for an explanation advantage in naive biological reasoning. *Cognitive Psychology*, *58*, 177–194. doi:10.1016/j.cogpsych.2008.06.002

Longbottom, S. E., & Slaughter, V. (2016). Direct experience with nature and the development of biological knowledge. *Early Education & Development*, *27*, 1145–1158. doi:10.1080/10409289.2016.1169822

Medin, D., Waxman, S., Woodring, J., & Washinawatok, K. (2010). Human-centeredness is not a universal feature of young children's reasoning: Culture and experience matter when reasoning about biological entities. *Cognitive Development*, *25*, 197–207. doi:10.1016/j.cogdev.2010.02.001

National Governors Association Center for Best Practices & Council of Chief State School Officers. (2010). *Common Core State Standards*. Washington, DC: Authors.

Ruckert, J. H. (2016). Generation conservation: Children's developing folkbiological and moral conceptions of protecting endangered species. *Early Education & Development*, *27*, 1130–1144. doi:10.1080/10409289.2016.1145005

Sarama, J., Brenneman, K., Clements, D. H., Duke, N. K., & Hemmeter, M. L. (2015). *Connect4Learning (C4L): The preschool curriculum*. Lewisville, NC: Gryphon House.

Shtulman, A., Neal, C., & Lindquist, G. (2016). Children's ability to learn evolutionary explanations for biological adaptation. *Early Education & Development*, *27*, 1222–1236. doi:10.1080/10409289.2016.1154418

Siegal, M., & Petersen, C. C. (1999). *Children's understanding of biology and health*. New York, NY: Cambridge University Press.

Solomon, G. E., & Zaitchik, D. (2012). Folkbiology. *Wiley Interdisciplinary Reviews: Cognitive Science*, *3*, 105–115.

Strouse, G. A., & Ganea, P. (2016). Are prompts provided by electronic books as effective for teaching preschoolers a biological concept as those provided by adults? *Early Education & Development*, *27*, 1190–1204. doi:10.1080/10409289.2016.1210457

Tanaka, F., Cicourel, A., & Movellan, J. R. (2007). Socialization between toddlers and robots at an early childhood education center. *Proceedings of the National Academy of Sciences, USA*, *104*, 17954–17958. doi:10.1073/pnas.0707769104

Tao, Y. (2016). Young Chinese children's justifications of plants as living things. *Early Education & Development*, *27*, 1159–1174. doi:10.1080/10409289.2016.1210456

Taverna, A. S., Medin, D. L., & Waxman, S. R. (2016). "Inhabitants of the earth": Reasoning about folkbiological concepts in Wichi children and adults. *Early Education & Development*, *27*, 1109–1129. doi:10.1080/10409289.2016.1168228

Thibaut, J.-P., Nguyen, S. P., & Murphy, G. L. (2016). Body and soul: Do children distinguish between foods when generalizing biological and psychological properties? *Early Education & Development*, *27*, 1250–1262. doi:10.1080/10409289.2016.1146551

Toyama, N. (2016). Preschool teachers' explanations for hygiene habits and young children's biological awareness of contamination. *Early Education & Development*, *27*, 38–53. doi:10.1080/10409289.2015.1036347

"Inhabitants of the Earth": Reasoning About Folkbiological Concepts in Wichi Children and Adults

Andrea S. Taverna, Douglas L. Medin, and Sandra R. Waxman

ABSTRACT

Across the world, people form folkbiological categories to capture their commonsense organization of the natural world. Structured in accordance with universal principles, folkbiological categories are also shaped by experience. Here we provide new evidence from the Wichi—an understudied indigenous community who live in the Chaco rainforest and speak their heritage language. A total of 44 Wichi (6- to 8-year-olds, 9- to 12-year-olds, adults) participated in an induction task designed to identify how broadly they attribute an invisible biological property (e.g., an internal organ) from 1 individual (either a human, nonhuman animal, or plant) to other humans, nonhuman animals, plants, natural kinds, and artifacts. *Research Findings*: These results (a) clarify the content of the Wichi's categories and the words they use to describe them, (b) showcase the power of covert (unnamed) categories, and (c) fortify the view that human-centered reasoning is not a universal starting point for reasoning about nature. *Practice or Policy*: Implications of these findings for early science education are discussed. In particular, we discuss (a) how the Wichi's construal of the natural world may be best integrated when they reach the (Western science–inspired) classroom and (b) how the current results bear on central issues in early science education more broadly.

Humans make their homes in nearly every corner of the earth, in the deserts of Africa, the tundra of the arctic, the farms of the Great Plains, and in inner cities like Chicago and Beijing. Children raised in any one of these communities will be surrounded by objects and events that children raised elsewhere may never encounter. Yet despite these striking differences in experiences, there are strong similarities in how people represent their natural habitats. People from across the world's communities form folkbiological categories. These folkbiological categories—a commonsense organization of the natural world—are structured in accordance with strong universal principles of hierarchical structure (Berlin, Breedlove, & Raven, 1973). More specifically, people across cultures tend to establish categories of objects that are nested within a hierarchical structure (e.g., TERRIER–DOG–ANIMAL). Moreover, within these structures, basic-level categories (those occupying an intermediate hierarchical position; e.g., DOG) have the strongest inductive force (Berlin et al., 1973; Rosch, Mervis, Gray, Johnson, & Boyes-Braem, 1976). Indeed, Berlin (1992) described basic-level categories as "beacons on the landscape" that serve to help people "carve the natural world at its joints" (p. 26). Of course, the content of the categories people form (e.g., basic-level categories like jaguars and chaguar in the Chaco rainforest or squirrels and tulips in Chicago) also bear the indelible stamp of experience within their surroundings.

This interplay between cultural universals and cultural specifics, so apparent in the folkbiological systems people create, has served as fertile ground for clarifying which aspects of human development might be universal and how these are shaped by experience. Adopting a cross-cultural and cross-linguistic developmental approach, psychologists have asked how young children acquire fundamental biological concepts, including ANIMAL, PLANT, and LIVING THING. Three powerful shaping forces have been identified: *language* (e.g., whether and how key folkbiological categories are marked in the native language; Anggoro, Waxman, & Medin, 2008; Berlin et al., 1973; Leddon, Waxman, & Medin, 2008; Leddon, Waxman, Medin, Bang, & Washinawatok, 2012; Taverna, Waxman, Medin, Moscoloni, & Peralta, 2014; Waxman, 2005), *forms of contact with the natural world* (e.g., the kinds of interactions people have with the different kinds [e.g., artifact-rich vs. artifact-sparse] of environments; Atran et al., 2001; Tarlowski, 2006; Taverna et al., 2014; Winkler-Rhoades, Medin, Waxman, Woodring, & Ross, 2010), and *culture* (e.g., community-wide belief systems about the natural world; Astuti, Solomon, & Carey, 2004; Atran & Medin, 2008; Taverna, Waxman, Medin, & Peralta, 2012; Waxman, Medin, & Ross, 2007).

Here we provide new evidence from another population—the Wichi, an understudied indigenous Amerindian group from a remote region of the Chaco forest (northern Argentina) who continue to speak their heritage language. Our work in this community is cross-disciplinary and cross-cultural at its core. Our team, which includes psychologists, linguists, and native members of the community, offers a rare opportunity to witness how folkbiological knowledge emerges in circumstances that differ rather dramatically from our own language, experiences of the natural world, and community-held belief systems. Perhaps more important, this work is also very timely. In 1984, the government of Argentina established an intercultural system of public schools aimed at providing the Wichi children with access to the Spanish language and to Western education in both the arts and sciences while preserving their cultural knowledge and native language ("Ley Argentina de Educación Nacional 26.206, 2006"). Unfortunately, however, curricular decisions have thus far been implemented by nonnative people (e.g., Zidarich, 2014). It is therefore important to identify what knowledge young Wichi children bring with them to their classrooms and to design curricula to augment this knowledge.

In the current article, we focus on folkbiological knowledge among the Wichi, asking how children and adults reason about the relations among living things and how their language and belief systems shape their systems of reasoning. We consider three distinct age groups, tapping into the knowledge in young children (to identify what knowledge they bring with them as they begin their education at 6 to 8 years of age), in older children (to identify their knowledge after a few years of schooling at 9 to 12 years of age), and in adults with little formal schooling (to identify how adults in the community reason about the natural world). This work, designed to broaden the empirical base in order to specify the universal and culture-specific processes underlying conceptual development, also has pressing implications for early science education, not only within the Wichi community but for education more broadly.

An Overview of The Wichi Community

This work is part of a broader investigation examining folkbiological knowledge in this indigenous Amerindian community. Although the Wichi have been described in the linguistics and anthropology literatures (e.g., Nercesian, 2011; Palmer, 2005; Suárez & Montani, 2010; Terraza, 2009; Vidal & Nercesian, 2009), to the best of our knowledge ours is the first investigation to have adopted a cognitive and developmental approach (Taverna et al., 2012, 2014).

In what follows, we describe in brief experiences in the natural world, Wichi beliefs, and the names the Wichi use to describe key folkbiological concepts.

Experience

The Wichi people are an indigenous population from the Chaco forest in the South American lowlands of Argentina. This region spans 1 million square kilometers and includes primarily grassy plains interspersed with distinct areas dominated by scrub growth, small woody plants, or palm groves. Our work takes place in the Wichi lawet community (Laguna Yema, Formosa), and in this community both children and adults have extensive and varied direct contact with different species of plants and animals, many of which have strong cultural functions and significance (Alvarsson, 1988; Arenas, 2003; Koschitzky, 1992; Maranta, 1987; Palmer, 2005; Suárez & Montani, 2010; Wilbert & Simoneau, 1982). The Wichi in this community depend economically on traditional activities, including hunting, fishing, and gathering; they engage in seasonal slash-and-burn horticultural practices. Textile weaving, pottery, and, to a lesser extent, ranching and farming are also important economic practices. Both women and men are engaged directly with plants and animals, but their activities reflect a strongly gender-based differentiation of labor and expertise. Men are responsible for hunting; fishing; and manufacturing wooden tools, furniture, and handicrafts. Women are responsible for gathering fruit, wood, and other plants, a task that typically requires full-day expeditions deep into the Chaco forest, on which they typically are accompanied by their children. In addition to going on these expeditions, children are engaged daily in activities embedded within the natural environment, including trapping lizards, walking in the forest, and swimming in the canal, among others (Taverna et al., 2012).

Beliefs

The cultural and religious beliefs held by the Wichi differ from those of most Western-oriented, technologically saturated communities. As in many other Amerindian communities, humans are seen as a part of—rather than apart from—the natural world (Bang, Medin, & Atran, 2007; Medin & Bang, 2014; Palmer, 2005; Pierotti, 2011; Wilbert & Simoneau, 1982). For example, there is a strong affinity between humans and nonhuman animals: In the Wichi origin myth, humans (males) emerged from the earth, climbing through a hole provided by the animals of the forest. It is interesting that because the first woman was pregnant and unable to pass through this hole, she arrived later climbing from a ladder in the sky. The Wichi believe that some humans and nonhuman animals have remained in the sky and that they become visible as the constellations of the night sky (Palmer, 2005; Wilbert & Simoneau, 1982). The narratives that support these beliefs—among many others—are of extraordinary importance and are frequently recounted in gatherings, spontaneous conversations (Wilbert & Simoneau, 1982), and school materials (Zidarich et al., 2006).

This close affinity between humans and nonhuman animals is also evident in the Wichi belief in *husek*, a concept that aligns roughly with Western notions of *spirit* or *soul*. Often identified with individual and collective goodwill (Palmer, 2005), *husek* is at the center of daily community life. The Wichi attribute *husek* equally to humans, nonhuman animals, and spiritual entities (*ahot*) but typically not to plants (Taverna et al., 2012).[1] The belief in *husek*, and the resulting close alignment between humans, nonhuman animals, and spiritual entities, links the biological and spiritual worlds: For both children and adults, spiritual entities, although invisible to the human eye, are not considered unreal or magical (Woolley, 1997) but instead are well-ensconced in folk theories of the natural world.

Language

In contrast to the languages of most Amerindian communities studied to date, the Wichi language is very much alive. Acquired naturally from infancy, it is the primary language within the family and in community life. Although some community members also speak some Spanish (especially schoolteachers and men doing commerce outside the community), Spanish is rarely used in conversations

[1]In Palmer's (2005) view, the term *husek* also has a secondary sense (vital will) that invokes a notion of vitality and may include certain species of plants imbued with magical properties (see Suárez & Montani, 2010, for a discussion); *husek* also appears to have a third sense (shamanistic will) that applies exclusively to a person who has become a shaman. This third sense was never mentioned in our interviews within the community (Taverna et al., 2012).

within the community. In our research, we focus on the Wichi folkbiologic lexicon rather than its grammar (but see Nercesian, 2011) because of strong cross-linguistic developmental evidence that naming and categories are intimately linked (Berlin et al., 1973; Waxman & Hall, 2004).

Why Naming Is Important

There is considerable evidence that even before infants utter their first words, language and conceptual development are linked: Naming supports the formation of object categories (see Waxman & Lidz, 2006, or Waxman & Gelman, 2009, for a review). For example, when infants observe a set of disparate members (e.g., dog, horse, duck) of a given object category (e.g., animal), they have difficulty noticing the category-based commonality among them. But when the very same members are introduced with a shared name, infants' categorization improves dramatically (Ferry, Hespos, & Waxman, 2010; Fulkerson & Waxman, 2007; Waxman & Markow, 1995).

This observation is directly relevant to cross-cultural and cross-linguistic work on folkbiology. After all, if naming supports object categorization in infants and young children, and if object categories serve as a basis for inductive inference, then the names children learn for biological entities should influence the categories they establish and their inductive strength. Recent experimental evidence has documented that language-specific differences in naming practices for biological entities do indeed influence children's concept acquisition (Anggoro, Medin, & Waxman, 2010; Anggoro et al., 2008; Hatano et al., 1993; Leddon et al., 2012; Stavy & Wax, 1989; Waxman, 2005; Winkler-Rhoades et al., 2010). For example, Anggoro et al. (2008) showed that 4- to 9-year-old English-speaking but not Indonesian-speaking children endorsed two different meanings of *animal*—ANIMAL$_{\text{INCLUSIVE}}$ and ANIMAL$_{\text{EXCLUSIVE}}$—and that this difference was mirrored in children's free sorting. When asked to identify entities that are "alive," older Indonesian-speaking children selected both plants and animals, but their English-speaking counterparts tended to exclude plants, which suggests that they may have misaligned *alive* with one of the *animal* senses (ANIMAL EXCLUSIVE).

The Wichi's Folkbiological Lexicon

To a Westerner's ear, this folkbiological lexicon is as surprising for the concepts that remain unnamed as for those that are named. See Figure 1 for a schematic representation of the Wichi folkbiological lexicon. The first thing to notice is the large number of categories that remain unnamed. According to Berlin (1992), among indigenous communities, it is not uncommon for higher order categories (which he described as "unique beginners" p. 15) to remain unnamed. In the anthropological and psychological literature, unnamed categories are described as *covert* categories (Berlin et al., 1973; Hays, 1976; Waxman, 2005).

Consider first the broad overarching folkbiological category that includes all living things. In English, this is described phrasally—*living things*. In Wichi, this is described using a compound noun—*hunhat lheley* (translated as "inhabitants of the earth")—whose scope includes not only all humans and nonhuman animals and plants but spiritual beings (see Taverna et al., 2012). Among these, only one category—*ahot* (spiritual beings)—is named.

The category corresponding to PLANTS remains unnamed in Wichi, but its constituent categories are named. Notice, though, that these named categories differ considerably in content from the taxonomic groupings named in most Western languages (e.g., trees vs. bushes). In Wichi, the named plant categories mark habitat-based groupings, including *hal'o* (wild trees and shrubs with woody trunks that inhabit the forest) and *tokos* (cultivated plants), among many others.

Similarly, although the category corresponding to (NONHUMAN) ANIMALS is unnamed, the Wichi lexicon includes superordinate names for *kinds* of animals. Again, these differ in content from the taxonomic groupings named in most Western languages (e.g., mammals vs. birds). In Wichi, the names mark habitat-based groupings, including *tshotoy* (animals of the forest), *tshotoy inot lheley* (animals of the water), *tshotoy fwiy'ohen* (animals of the air), and others, including *laloy* (domestic animals). Although there is strong consensus that this is the primary meaning of *tshotoy*, there is

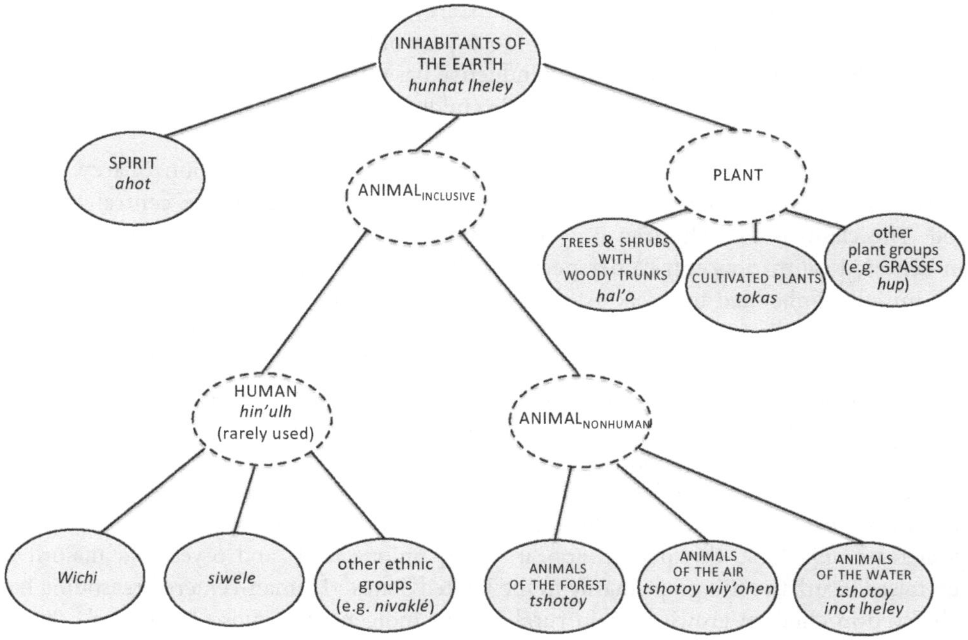

Figure 1. A schematic depiction of Wichi names for fundamental folkbiological concepts. Note that the nodes corresponding to $\text{ANIMAL}_{\text{INCLUSIVE}}$ (human and nonhuman combined), $\text{ANIMAL}_{\text{NONHUMAN}}$, and PLANT are unnamed.

considerable debate about whether *tshotoy* can also refer (more inclusively) to all nonhuman animals. Some suggest that in certain conversational contexts it can; others argue strenuously that *tshotoy* cannot be used to refer to any animals other than the animals of the forest (Taverna et al., 2012; see Berlin et al., 1973, for a discussion of the phenomenon in which the same word may be used for both a superordinate category and a kind within it).

Finally, consider how the Wichi refer to humans. There is a word that refers to all humans (*hin'uhl*), but it is very seldom used. Instead, the Wichi favor more specific terms that mark community affiliation: *Wichi* (Wichi people), *siwele* (White, non-Wichi people), and a number of other nouns that apply to people of other cultural groups (e.g., *nivaklé*).

This descriptive information about the community, its beliefs, and its language provides the backdrop for our investigations of folkbiological reasoning in Wichi children and adults.

Tapping Into Wichi Folkbiological Knowledge

What is the relation between humans, nonhuman animals, and plants in Wichi child and adult reasoning about the natural world? To tap into their reasoning, we used a *category-based induction* task, one that has been instrumental in investigations of biological reasoning in children and adults from a range of communities, from urban to rural communities, from majority-culture to indigenous communities (e.g., Anggoro et al., 2010; Atran et al., 2001; Carey, 1985; Gelman, 1988; Herrmann, Waxman, & Medin, 2010; Ross, Medin, Coley, & Atran, 2003; Tarlowski, 2006; Waxman et al., 2007). In this task, participants are introduced to a novel property of an entity (the *base*) and then asked whether this property can be generalized to other entities (the *targets*). For example, they may be taught that dogs have a novel biological property (e.g., an omentum) and asked whether other entities (typically including a range of animals, plants, and artifacts) also share this property. These attributions—or generalizations—of the novel biological property shed light on how people construe the relations among entities in the natural world, especially the relation between human and nonhuman animals.

In her pioneering work with this task, Susan Carey (1985) argued that in reasoning about the natural world (a) young children (roughly 4- to 6-year-olds) universally adopt an anthropocentric stance in which humans serve as a stronger inductive base than do nonhuman animals and that (b) this human-centered stance is replaced in middle childhood as children begin to realize that humans are one kind of animal among many.

Although over the years Carey has amended the details of her position (Carey, 1995, 1999; Carey, Zaitchik, & Bascandziev, 2015), to the best of our knowledge the central tenet has not changed. The claim is that children universally begin reasoning about the natural world from a human-centered (anthropocentric) perspective and that their initial understanding of the natural world is initially embedded in a *folk psychological* rather than a *folk biological* explanatory framework. Nonetheless, there have been challenges to this claim documenting, for example, that preschool-age children distinguish biological and psychological construals of living things (Coley, 1995; Inagaki & Hatano, 1993) and use explicitly biological causal mechanisms to explain biological phenomena (e.g., Rosengren, Gelman, Kalish, & McCormick, 1991; Solomon & Cassamites, 1999; Springer, Ngyuen, & Samaniego, 1996).

In addition, cross-cultural studies based on populations that differ in important ways from the ones observed by Carey have overturned this notion of a universal anthropocentric starting point for biological reasoning. First, this pattern appears to be unique to 4- and 6-year-old majority-culture children raised in urban settings, primarily in the United States[2]: Human-centered reasoning has never been observed in children growing up in rural U.S. or indigenous groups (Atran et al., 2001; Coley, Medin, & James, 1999; Herrmann, Medin, & Waxman, 2012; Inagaki, 1990; Ross et al., 2003). Second, even within this population, human-centered reasoning is not the starting point for reasoning about the biological world (Herrmann et al., 2010; Waxman, Herrmann, Woodring, & Medin, 2014). These studies documented that children younger than 5 years of age are decidedly not anthropocentric in their reasoning. Therefore, human-centered reasoning is not a universal starting point but is instead an acquired perspective, supported (however briefly) in some, but by no means all, communities. We have proposed that this anthropocentric stance in urban-raised children reflects their relative paucity of direct contact with other species of natural world, coupled with the frequent and powerfully engaging anthropocentric images in children's books, movies, and other media (Herrmann et al., 2010).

What Can Be Learned From Using Category-Based Induction to Tap Into Folkbiological Knowledge Among Wichi Children and Adults?

We focus on four key issues.

(1) What is the inductive status of the (covert) category NONHUMAN ANIMAL? Although unnamed in Wichi, does this category nonetheless function as a coherent, albeit covert, base for reasoning? Or do the named habitat-based categories (e.g., forest, water, air) provide distinct inductive power?

(2) What is the inductive status of the category ANIMAL (including HUMAN AND NONHUMAN ANIMALS)? Although no single word refers to both humans and nonhuman animals (Taverna et al., 2014), there is nevertheless a close affinity between humans and nonhuman animals. We therefore suspect that the Wichi represent a covert category—encompassing humans and nonhuman animals—that guides their reasoning about the biological world.

[2] In the original design, Carey presented children with only a single base—either a dog or a human. Some subsequent studies have maintained this between-participants design, but others (Anggoro et al., 2010; Atran et al., 2001; Ross et al., 2003) have presented more than a single base to each child. Close analysis of these studies reveals that the order in which the bases are presented has consequences for young urban children's patterns of performance. Anggoro et al. (2010) documented base order effects in young urban children from Chicago and Jakarta: If a nonhuman animal serves as the child's first base, the child's tendency to extend the novel biological property to the (subsequent) human base is much weaker. See Medin et al. (2010) and Anggoro et al. (2010) for a discussion.

(3) How do the Wichi construe the relation between humans and nonhuman animals? The Wichi's extensive direct experience with the natural world, coupled with the absence of anthropocentric media images for children, led us to predict that even the youngest Wichi children in our investigation would not adopt a human-centered (anthropocentric) stance but would instead view humans and nonhuman animals as comparable in inductive strength.

(4) How do plants fit in with the Wichi's reasoning about the natural world? Because the Wichi have such extensive firsthand experience with plants and accord them strong cultural significance, we expected the category PLANT to have strong inductive power. At issue is the inductive status of this (covert) category. Do the Wichi reason about plants as a cohesive category, even though it is unnamed? Although we included fewer plants than animals in the induction task, the results should nonetheless offer initial insight into how the Wichi extend novel biological properties both to and from members of this unnamed category.

Method

Participants

A total of 44 native speakers of Wichi participated: 16 children ages 6 to 8 years old (nine females, *M* age = 7.1), 16 children ages 9 to 12 years old (seven females, *M* age = 10.6), and 12 adults (10 females, *M* age = 32.0, range = 18–49 years). The participants, who spoke exclusively Wichi in the home and had very limited knowledge of Spanish, were interviewed exclusively in Wichi. All children attended the Wichi *Lako* School, a public school in Laguna Yema, Formosa (Argentina). This is one of several Wichi schools implemented by the Argentine government in 1984 as part of the Intercultural Education Program. The teachers in these schools include both native Wichi teachers who are bilingual in Spanish and nonnative teachers who are monolingual speakers of Spanish. Although some Spanish is introduced (lightly) in the first grade (at roughly age 6), children communicate almost exclusively in Wichi in school until the later grades (at roughly age 9 or 10); they overwhelmingly use Wichi outside of school. As for the adults, most had limited contact with Western education: Six had not completed elementary school, four had completed elementary school, and only two had completed secondary school.

Materials

Images

Eighteen colored photographs depicting a range of living and nonliving entities served as visual stimuli. Four of the entities served as bases (human, jaguar, dog, carob tree); the remaining 14 photographs served as targets (see Table 1). We selected items that were familiar to both children and adults in the Wichi community. Each photograph was presented on an 8.5 × 5.5" laminated card.

One other aspect of our stimulus selection bears mention: Because most of our questions concerned the Wichi's interpretation of the relation among animals—both human and nonhuman animals—and their inductive status, it was essential that we include animals from each of the habitat-based named groupings. This demand, coupled with our decision to keep the number of targets and bases comparable to other closely related work with adults and children, meant that we presented more animals than plants. Nonetheless, we included one bush and one tree among our plant targets in an effort to gather preliminary insight into whether and when the Wichi treat plants as a cohesive category with inductive potential.

Property Names

Working in collaboration with several adult native Wichi speakers, we created four novel words (*t'i, l'hele, lachuwej, lawit'ui*) to be used as names for the novel properties in the induction task (described in "Procedure").

Table 1. Complete List of Items Used as Bases and Targets.

Bases	*Targets*
Human	Human
Jaguar	Anteater[a]
Dog	Turtle[a]
Carob tree	Worm[a]
	Caracara[b]
	Parrot[b]
	Fly[b]
	Cayman[c]
	Piranha[c]
	Hardwood tree
	Acacia bush
	Water
	Chair
	Bicycle

[a]Animals of the forest. [b]Animals of the air. [c]Animals of the water.

Procedure

Participants were interviewed individually in Wichi, sitting across from the experimenter in a quiet room. Children were interviewed in a quiet section of a community center; adults were interviewed in their homes. Before the experiment began, children participated in a warm-up task; after the experiment ended, adults were asked to provide justifications for their responses. Each participant completed the induction task four times, each time with the same targets but a different base and novel property. The order in which bases were presented was randomized across participants. Here we use the jaguar base to illustrate.

Warm-Up Task (Children Only)

For children, the experimenter began the session with a brief warm-up task designed to clarify that it was acceptable to answer either "yes" or "no" to the questions she would be posing. For example, the experimenter showed the child a line drawing of a shape (e.g., a red triangle), asking, for example, "Is this red? Is it a square?" All children readily answered "yes" and "no" and did so correctly.

Induction Task (Children and Adults)

All participants completed the induction task across four trials, each trial using a different base; bases were presented in one of four Latinized orders. Before each trial began, the targets were shuffled and presented in random order.

Training phase. To begin, the experimenter presented one base (e.g., a photograph of a jaguar) and asked participants to name it. If they named it correctly, they were given positive feedback. If not, they were gently corrected. Next the experimenter handed the participant a line drawing of the base (e.g., a jaguar) and took one for herself as well. She then introduced a novel biological property (e.g., "Jaguars have *t'i* inside them. *T'i* is a greenish substance, and it goes inside!"). She then handed the participant a crayon, saying, "Look! I'm drawing *t'i* in my picture of a jaguar! Will you draw *t'i* in yours?" She then gathered the line drawings and moved on to the test phase.

Identifying the target cards. Before beginning the initial test (the first of four), participants participated in a naming event designed to ensure their familiarity with the depicted entities. The experimenter revealed each of the photographs, in random order, asking the participant to identify each by name, then providing feedback. If the participant named an entity incorrectly, the experimenter supplied the correct name by saying, for example, "It may *look* like a [X], but it's actually a [Y]." At this point, the induction task began. This naming event was not repeated on the three subsequent induction tasks.

Test phase. The experimenter shuffled (randomized) the target photographs, saying, for example, "Remember when we talked about *t'i?* And we said that jaguars have *t'i* inside? Some other things have *t'i* too. Please tell me: Which of these [pointing to her pile of target photographs] have *t'i* inside too, okay?" At this point, she revealed each target sequentially, in random order, asking, "What do you think? Do [Xs] have *t'i* inside, like jaguars do?" She recorded the participant's response and then moved on to the next target, and so on. Questions were phrased in the generic, focusing on the kinds in question ("Do anteaters have *t'i* inside, like jaguars do?") rather than on the particular individuals depicted ("Does this anteater have *t'i* inside, like this jaguar does?").

Participants then completed the training and test phases for the remaining three bases; a different novel property was introduced for each base.

Justifications (Adults Only)

After completing the induction task for all four bases, adults were asked to justify their responses on the final round. The experimenter said, for example, "You told me that these [pointing to the entities to which the participant attributed the novel property] have *t'i* like jaguars do. Why do you think that is so?"

Results

Before analyzing the results formally, we sought some assurances that this task was a suitable one for tapping into the Wichi children's and adults' reasoning about the natural world. We reasoned as follows: If the Wichi remember the novel properties, and if they are willing to judge the likelihood that various other entities might share these properties, then they should provide a systematic pattern of responses as a function of the similarity of the base to the target categories or in terms of Figure 1, as a function of distance within the hierarchy. An examination of Table 2, which presents the proportion of generalizations from each base to each of the targets for each age group, suggests that they were indeed systematic.

Consider, for example, the Wichi's responses to the artifact targets. If this task did indeed tap into the Wichi's considerable knowledge about the natural world, then they should have rarely extended a novel biological property, introduced in conjunction with a biological entity (any of our four bases), to an artifact (chair, bicycle). A glance at Table 2 reveals that they rarely extended novel properties to the manmade artifacts at any age. Consider next their responses to the human target. If the Wichi were responding systematically to this task, then they should have been more likely to extend a novel property to the human target if the property was introduced on the human base than on a nonhuman base (either the jaguar, the dog, or the carob tree). Table 2 reveals that extensions to the human target were uniformly high when the novel property was introduced on a human base and attenuated when it was introduced on a nonhuman base.

Taken together, these observations provide assurances that the current paradigm tapped into a systematic set of inferences about the natural world among Wichi children and adults. With this as a foundation, we go on to consider more precisely the Wichi's patterns of inductive inference.

What Is the Inductive Status of the (Covert) Category NONHUMAN ANIMAL?

In reasoning about entities in the natural world, do the habitat-based distinctions among the named animal kinds (*tshotoy, tshotoy wiy'ohen, tshotoy inot lheley*) hold distinct inductive power? Or does participants' reasoning reveal the inductive power of the more inclusive, though by most accounts covert, category NONHUMAN ANIMAL?

To test this, we calculated each participant's mean response to three animal categories—animals of the forest (anteater, turtle, worm), animals of the air (caracara, parrot, fly), and animals of the water (caiman, piranha)—and submitted these animal category scores to an analysis of variance (ANOVA) using base (4: human, jaguar, dog, carob tree) and animal target category (3: forest, air, water animal) as within-participants factors and age (3: 6- to 8-year-olds, 9- to 12-year-olds, adults) as a between-participants factor (see Figure 2). The ANOVA revealed a main effect for base, *F*(3,

Table 2. Means (*SD*) for Generalizations From Each Base to Each Target at Each Age.

		Animals of the Forest			*Animals of the Air*			*Animals of the Water*		*Plants*			*Artifacts*	
Base	*Human*	*Anteater*	*Turtle*	*Worm*	*Caracara*	*Parrot*	*Fly*	*Caiman*	*Piranha*	*Hardwood*	*Acacia*	*Water*	*Chair*	*Bicycle*
6- to 8-year-olds														
Human	0.87 (0.34)	0.5 (0.51)	0.44 (0.51)	0.63 (0.5)	0.63 (0.5)	0.44 (0.51)	0.44 (0.51)	0.63 (0.5)	0.50 (0.51)	0.31 (0.47)	0.25 (0.44)	0.25 (0.44)	0.06 (0.25)	0.19 (0.4)
Jaguar	0.75 (0.57)	0.75 (0.44)	0.56 (0.51)	0.75 (0.44)	0.81 (0.4)	0.75 (0.44)	0.87 (0.34)	0.87 (0.34)	0.94 (0.25)	0.50 (0.51)	0.37 (0.50)	0.25 (0.44)	0.19 (0.40)	0.19 (0.4)
Dog	0.56 (0.51)	0.75 (0.44)	0.75 (0.44)	0.69 (0.47)	0.63 (0.5)	0.56 (0.51)	0.75 (0.44)	0.81 (0.4)	0.75 (0.44)	0.31 (0.47)	0.25 (0.44)	0.13 (0.34)	0.13 (0.34)	0 (0)
Carob tree	0.60 (0.49)	0.47 (0.50)	0.40 (0.49)	0.53 (0.50)	0.60 (0.49)	0.40 (0.49)	0.53 (0.50)	0.60 (0.49)	0.40 (0.49)	0.93 (0.25)	0.53 (0.50)	0.33 (0.47)	0.20 (0.40)	0.13 (0.34)
9- to 12-year-olds														
Human	0.94 (0.25)	0.81 (0.40)	0.75 (0.44)	0.50 (0.51)	0.75 (0.44)	0.69 (0.47)	0.69 (0.47)	0.69 (0.47)	0.69 (0.47)	0.44 (0.51)	0.19 (0.40)	0.31 (0.47)	0 (0)	0 (0)
Jaguar	0.56 (0.51)	0.94 (0.25)	0.75 (0.44)	0.81 (0.40)	0.75 (0.44)	0.81 (0.4)	0.63 (0.50)	0.81 (0.40)	0.81 (0.40)	0.38 (0.50)	0.13 (0.34)	0.13 (0.34)	0.13 (0.34)	0 (0)
Dog	0.69 (0.47)	0.75 (0.44)	0.50 (0.51)	0.56 (0.51)	0.69 (0.47)	0.50 (0.51)	0.38 (0.50)	0.63 (0.50)	0.44 (0.51)	0.25 (0.44)	0 (0)	0.13 (0.34)	0 (0)	0 (0)
Carob tree	0.44 (0.51)	0.37 (0.50)	0.31 (0.47)	0.44 (0.51)	0.50 (0.51)	0.37 (0.50)	0.44 (0.51)	0.38 (0.50)	0.44 (0.51)	0.94 (0.25)	0.56 (0.51)	0.13 (0.34)	0.13 (0.34)	0 (0)
Adults														
Human	0.92 (0.28)	0.25 (0.45)	0.17 (0.38)	0.17 (0.38)	0.25 (0.45)	0.33 (0.49)	0.17 (0.38)	0.50 (0.52)	0.33 (0.49)	0.17 (0.38)	0.17 (0.38)	0.08 (0.28)	0 (0)	0 (0)
Jaguar	0.58 (0.51)	0.83 (0.38)	0.33 (0.49)	0.42 (0.51)	0.75 (0.45)	0.33 (0.49)	0.33 (0.49)	0.58 (0.51)	0.33 (0.49)	0.25 (0.45)	0.25 (0.45)	0 (0)	0 (0)	0 (0)
Dog	0.50 (0.52)	0.67 (0.49)	0.42 (0.51)	0.25 (0.45)	0.58 (0.51)	0.25 (0.45)	0.33 (0.49)	0.58 (0.51)	0.33 (0.49)	0.17 (0.38)	0.17 (0.38)	0 (0)	0 (0)	0 (0)
Carob tree	0.17 (0.38)	0.17 (0.38)	0.17 (0.38)	0.17 (0.38)	0.08 (0.28)	0.08 (0.28)	0.08 (0.28)	0.17 (0.38)	0.17 (0.38)	0.92 (0.28)	0.75 (0.45)	0.33 (0.49)	0.08 (0.28)	0 (0)

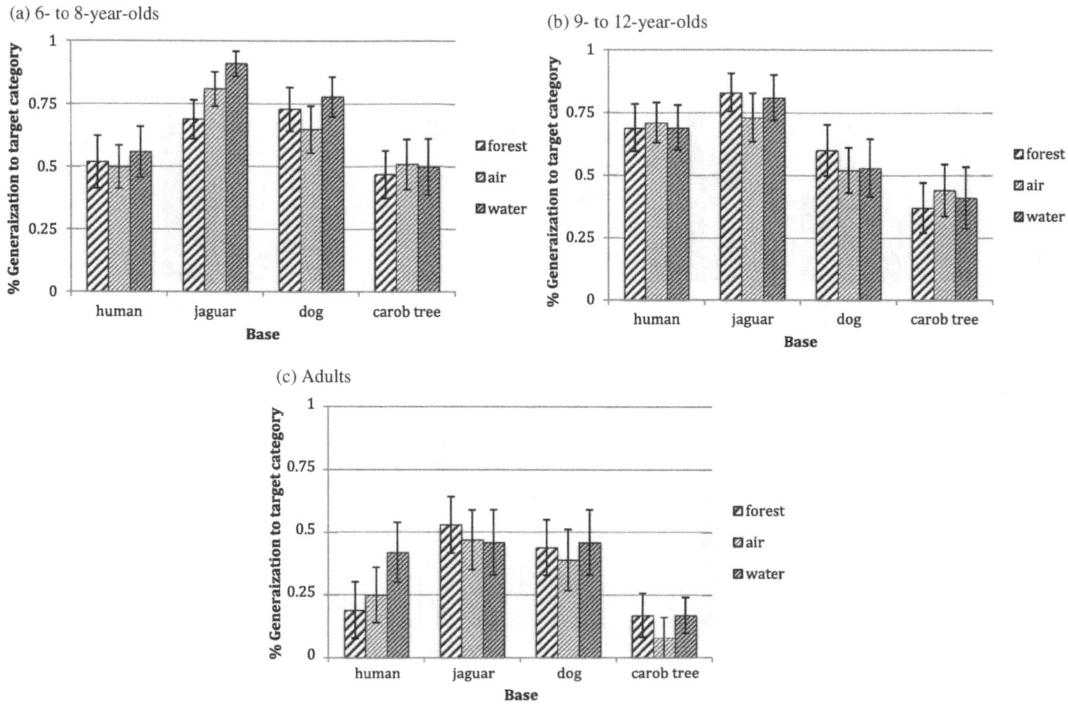

Figure 2. Mean generalizations (and standard deviations) from each base to animal target groups as a function of age.

120) = 12.026, $p < .001$; participants were more likely to project the novel biological property to the animal targets from the jaguar base than from the remaining three bases. There was also a main effect for age, $F(2, 40) = 5.585$, $p = .007$. There were no other main effects or interactions.

This analysis reveals that although the Wichi certainly identify and name distinct habitat-based categories of animals, when it comes to reasoning about the natural world, the covert category nonhuman ANIMAL serves as a coherent basis for induction.

Based on this outcome, in all subsequent analyses we combined participants' responses involving nonhuman animals (either as targets or as bases), treating animals as a single group (see Figure 3). We first submitted these data to an ANOVA with base (3: human, nonhuman animal, plant) and target category (5: human, nonhuman animal, plant, water, artifact) as within-participants factors and age (3: 6- to 8-year-olds, 9- to 12-year-olds, adults) as a between-participants factor. This comprehensive ANOVA revealed a main effect for target category, $F(4, 328) = 70.029$, $p < .001$. This was qualified by an interaction between target category and base, $F(8, 328) = 23.821$, $p < .001$; participants at all ages were more likely to project a novel property to the plant targets when it was introduced on the carob tree than on the human or nonhuman animal bases. The comprehensive ANOVA also revealed a main effect of age, $F(2, 41) = 4.313$, $p < .05$; adults were more conservative in projecting novel properties than were children at both ages.

In all remaining analyses, we used this comprehensive analysis as a foundation for conducting planned comparisons at each age to address the remaining three questions specifically. For all comparisons, $p < .05$ was set as the threshold for statistical significance.

What Is the Inductive Status of the Category ANIMAL *(Including* HUMAN AND NONHUMAN ANIMALS*)?*

An examination of the Wichi's response patterns at every age offers three strong indications that they use a covert category that includes both humans and nonhuman animals (analogous in scope to the English animal-inclusive) in reasoning about the biological world. More specifically, at every age,

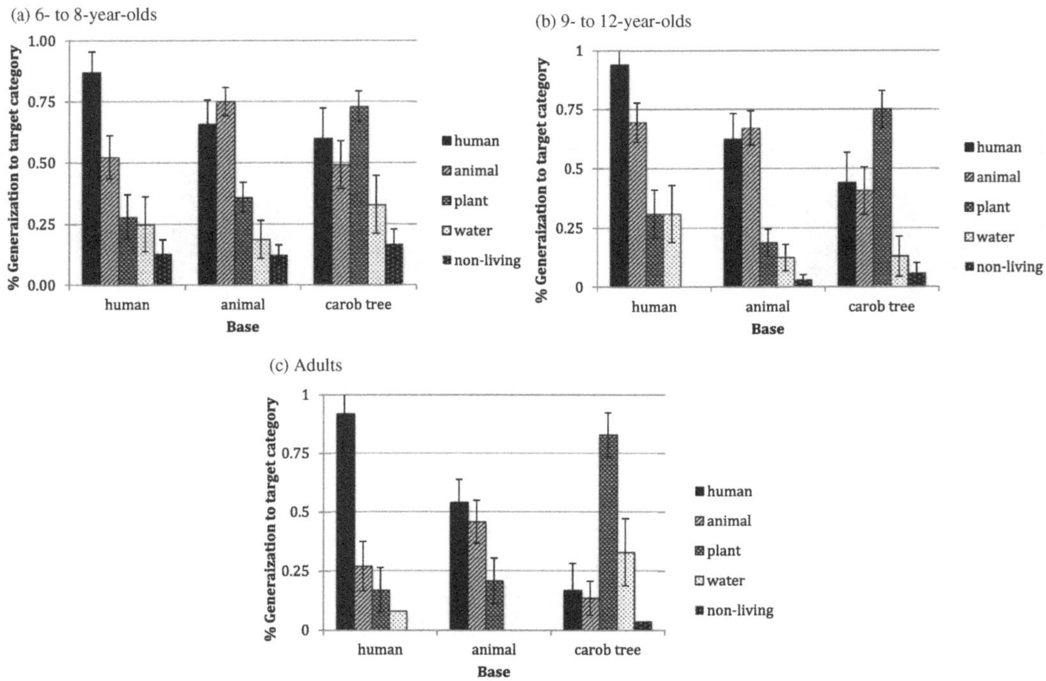

Figure 3. Mean generalizations from human, animal, and carob tree bases as a function of age.

the Wichi were (a) more likely to extend a novel property taught on a nonhuman animal to a human than to plants (all *ps* < .05), (b) more likely to extend a novel property taught on a human to nonhuman animals than to plants (all *ps* < .05),[3] and (c) equally likely to extend a novel property from a plant to either a human or a nonhuman animal.

Thus, although the category that includes both human and nonhuman animals remains unnamed in Wichi, children and adults alike appreciate the close affinity between them, suggesting the use of a covert category that includes both humans and nonhuman animals as a basis for reasoning about the biological world.

How Do the Wichi Construe the Relation Between Humans and Nonhuman Animals?

We address this question by focusing our analysis on the two patterns of inductive reasoning that have served as signatures of anthropocentric reasoning in past research (Anggoro et al., 2010; Astuti et al., 2004; Carey, 1985; Herrmann et al., 2010; Ross et al., 2003). More specifically, we considered whether participants at each age were (a) more willing to project the novel property from the human base to nonhuman target animals than from the nonhuman animal bases to human targets and (b) more willing to project the novel property from the human base to nonhuman target animals than from the nonhuman animal bases to nonhuman target animals. We found no evidence of human-centered reasoning among the Wichi in our sample at any age.

6- to 8-Year-Olds

First, our youngest participants were just as likely to extend a novel property from a human to a nonhuman animal as from a nonhuman animal to a human ($M_{6\text{- to 8-year-olds}}$ = .523 and .656,

[3]There was only a single exception to this pattern: Among adults, among which generalizations from humans were exceedingly low, the difference between their extensions from a human to nonhuman animals (.271) versus from a human to a plant (.167) was in the predicted direction but was not statistically significant.

respectively; $p = .25$). Second, they were significantly more likely to extend a property from a nonhuman animal to other nonhuman animals (.750) than from a human to a nonhuman animal (.523; $p < .05$). Notice then that these children actually favored nonhuman animals over humans as an inductive base; this is the opposite of a human-centered pattern of reasoning.

9- to 12-Year-Olds

The older children also were just as likely to extend a novel property from a human to a nonhuman animal as from a nonhuman animal to a human ($M_{9\text{- to 12-year-olds}} = .695$ and .625, respectively; $p = .415$). They were also equally likely to extend a property from a nonhuman animal to other nonhuman animal (.672) as from a nonhuman animal to a human (.625; $p = .753$).

Adults

For the adults, nonhuman animals provided a significantly stronger inductive base than did humans. Adults were significantly more likely to extend a novel property from a nonhuman animal to a human than from a human to nonhuman animals ($M_{\text{adults}} = .542$ and .271, respectively; $p < .05$). In addition, they were significantly more likely to extend a property from a nonhuman animal to other nonhuman animals (.458) than from a human to a nonhuman animal (.271; $p < .05$).

In sum, these analyses suggest that among the Wichi, humans do not serve as a privileged inductive base in reasoning about the animal kingdom. In the next analysis, we considered the responses of each individual participant, rather than group means, to assess whether this decidedly non-anthropocentric pattern of reasoning, evident in the mean analyses (described previously), is characteristic of most individuals. To address this question, we tallied each individual's response for each of the two signature comparisons ([a] and [b] above).

For (a), for each participant, we compared the number of projections from the human base to nonhuman target animals against the number of projections from the nonhuman animal bases to human targets. Each participant was coded either as *anthropocentric* (if the number of projections from the human base to nonhuman target animals exceeded the number of projections from the nonhuman animal bases to human targets) or *not anthropocentric* (if the number of projections from the human base to nonhuman target animals was less than or equal to the number of projections from nonhuman animal bases to human targets).

For (b), for each participant, we compared the number of projections from the human base to nonhuman target animals against the number of projections from nonhuman animal bases to the nonhuman target animals. Each participant was coded either as *anthropocentric* (if the number of projections from the human to nonhuman animals exceeded the number projections from nonhuman animals to nonhuman animals) or *not anthropocentric* (if the number of projections from the human to nonhuman animals was less than or equal to the number of projections from nonhuman animals to nonhuman animals).

Participants' responses for each comparison are reported in Table 3. At every age, the number of non-anthropocentric responses outnumbered the number of anthropocentric response by a factor of at least 2 and in most cases by a factor of at least 4.

Following Carey (1985) and Herrmann et al. (2010), we reasoned that the best evidence for anthropocentric reasoning requires an analysis of participants' responses to both comparisons (a) and (b). We therefore tabulated the number of participants at each age who responded in a consistently anthropocentric fashion to both comparisons, responded in a consistently non-anthropocentric fashion to both comparisons, and responded inconsistently (i.e., by providing one anthropocentric and one non-anthropocentric response). As can be seen in Table 4, consistently anthropocentric patterns of response were rare at every age. Moreover, children at both ages were significantly more likely to exhibit a consistently non-anthropocentric than consistently anthropocentric pattern: 6- to 8-year-olds, $\chi^2(1, N=13) = 6.231$, $p = .01$; 9- to 12-year-olds,

Table 3. Individual Participants' Responses at Each Age to the Two Signature Indices of Human-Centered Reasoning.

	(a) (Human:Nonhuman Animal) > (Nonhuman Animal: Human)			*(b) (Human: Nonhuman Animal) > (Nonhuman Animal: Nonhuman Animal)*		
Age Group	*Anthropocentric*	*Non-Anthropocentric*	*Total*	*Anthropocentric*	*Non-Anthropocentric*	*Total*
6 to 8 years	3	13	16	2	14	16
9 to 11 years	5	11	16	3	13	16
Adults	0	12	12	0	12	12

$\chi^2(1, N = 11) = 7.364$, $p = .007$.[4] Not a single adult exhibited a consistently human-centered pattern of reasoning about the natural world.

Clearly then, analyses based on group means and on individual participants' responses converged to suggest that the Wichi do not privilege humans as an inductive base. Moreover, in this population, nonhuman animals appear to be a stronger base for induction to other nonhuman animals than are humans.[5] This outcome suggests that even among the youngest children in our sample,[6] humans are represented as one animal among many, and this despite the fact that the category including human and nonhuman animals is not named.

How Do Plants Fit in With the Wichi's Reasoning About the Natural World?

Although we included fewer plants than animals, the results nonetheless provide a glimpse of Wichi reasoning about the plant kingdom at each age. Planned comparisons revealed that for children and adults alike, the Wichi were significantly more likely to extend a novel biological property to the plant targets if it was introduced on the carob tree than on either the human or nonhuman animal base (all *ps* $< .05$).

Adult Justifications

The adults' justifications, provided in Table 5, offer a different vantage point into Wichi reasoning about the natural world. Fortunately, adults offered justifications on 71% of their opportunities to do so (on 34 of 48 possible [4 bases × 12 adult participants] trials).

Table 4. Number of Individual Participants at Each Age and Their Responses to the Two Signature Indices of Human-Centered Reasoning.

Age Group	*Consistently Anthropocentric (on Both [a] + [b])*	*Consistently Non-Anthropocentric (on Neither [a] nor [b])*	*Inconsistent*	*Total*
6 to 8 years	2	1 1	3	16
9 to 11 years	1	10	5	16
Adults	0	12	0	12

Note. See Table 3 and "How Do the Wichi Construe the Relation Between Humans and Nonhuman Animals?" for descriptions of (a) and (b).

[4]Notice that this analysis is a conservative one; we excluded from the analysis the few children exhibiting inconsistent responses. (Including these children in the analysis did not change the results.)

[5]An analysis of order effects provided even more compelling evidence that the Wichi do not favor humans over nonhuman animals as an inductive base. In prior work, we documented that in populations in which young Western-educated children showed human-centered asymmetries (cf. Chicago, Jakarta), this anthropocentric pattern was most pronounced among children for whom a human served as the first base compared to those who got the human base later (Anggoro et al., 2010). To ascertain whether there was any hint of anthropocentrism among the Wichi, we asked whether anthropocentric reasoning was evident in participants who got human as their first base. There was no hint of human-centered reasoning whether the human base was first or last.

[6]Even among 6-year-olds, the age group revealing anthropocentric patterns of reasoning in Carey (1985), only a single child consistently favored the human over the nonhuman animal bases. All five of the remaining 6-year-olds were consistently non-anthropocentric.

Table 5. Adult Justifications.

Human Base

H1. "because all of them are animals *(tshotoy)*, all them can move away and run, if there is something wrong or bad they can run away in contrast to these ones (*gestures to plants*) which they cannot"

H2. "because all of them have husek and can breathe as humans (*hin'ulh*)

However, they (gestures to human and trees) have the same inside ... the human (*siwele*) has the blood that runs inside and trees have a green sap, the greenery, it is the same ...I have this thought ...the person () is different than the trees (*hal'o*) because he feels ... and can be moved if there is danger he can run away ... trees (*hal'o*) can not and that makes it more helpless ..."

H3. "because they eat fruits from the forest, they eat the same"

H4. "because all of them are large"

H5. "because the way they eat, they eat meat"

H6. "because humans () and trees have the same body, are equal, are high and they are standing. At the top are the same, have the foliage which makes them similar, it protects them, and when each foliage grows it is never the same as others. Each person (hin'u) has their own and each tree (*hal'o*) has its own. Never there are two "mistoles" or carob trees alike. They are different because they grow up differently ... the place where it is, one is more sheltered and better ..."

Jaguar Base

J1. "because all of them have *husek* and are in the land"

J2. "all of them are like the jaguar, the jaguar has a lot of contact with the "pachamama" it is special, it has a lord who is a *duende de monte* and it protects the jaguar"

J3. "because all are animals (*tshotoy*) they move by themselves and are alive"

J4. "because they both are on the forest (*monte*) they come from the same place. That is why the have the same substance inside...the caiman is also in the forest but it spends much time on the river which is a very different place to the forest, both places are different with very different inhabitants (*lheley*) "

J5. "because all them eat the same, have the same substance inside, and all of them are big"

J6. "because they eat the same, meat"

J7. "because all them eat the same, have the same substance inside, and all of them are big"

Dog Base

D1. "because all of them are from the land"

D2."because all of them are animals (*tshotoy*), they eat and are alive"

D3. "because all of them eat meat"

D4. "because all them are big/large animals (*tshotoy*) that is why they have the same substance inside"

D5. "because the dog and the human (hin'u) are in the same place, they do the same and they are together"

Carob Tree Base

C1. "because all them are in the land, in contrast I think the piranha is not like the carob tree and it has not the same substance inside because the piranha is in the river and it is very different than these ones (*gestures to the land animals*). But the piranha has the same inside than the water because she needs it to be alive."

C2. "because they are not animals (*tshotoy*), even though the carob tree is like an animal (*animale*, borrowed from Spanish) they feed us"

C3. "because they both grow up and can die and when we go to the forest and look for the algarroba (carob tree fruit) sometimes we want to pick it up and it refuses, so something inside prevents you not to pick it up because it may damage you, or hurt you"

C4. "because they are green, they give fruits, and the trees (*hal'o*) have the greenness inside, they are in the forest, they grow up there and they stay" there and all of them have the same liquid inside. The carob tree has a very hard bark and it is used for the leather"

C5. "because all of them need themselves that is why they have the same substance"

C6. "because all of them need each other... all need water and that is why they have the same substance"

C7. "the parrot eats the leaves and that is why they can have the same substance inside them. The caiman, water and plants (*plantas*, borrowed from Spanish) also have the same because all need water to live; besides the trees and the caiman have the same substance that is why they both dry when they die". The carob tree is holy because it feeds us every single year"

C8. "because they have the same body, the head and the feet, both are equal and grow up in the same manner"

An inspection of Table 5 reveals that in their justifications, the Wichi frequently invoked taxonomic relations or shared morphology (e.g., J3: "because all are animals [*thsotoy*], they move by themselves and are alive"), either alone or in combination with other more ecological and relational explanations, including explicit mention of shared habitats (e.g., J4: "they are both in the forest ... come from the same place"), shared behaviors (e.g., H3: "they eat fruits of the forest"), animacy status (e.g., H1: "all of them can move away and run"), or spiritual status (e.g., "because all of them have *husek* ...").

A covert category of animals? Their performance in the induction task suggested that the Wichi use the covert category ANIMAL in reasoning about the natural world. The adults' justifications offer converging evidence for this interpretation. On only four occasions did an adult make reference to an inclusive category ANIMAL. On three such occasions, *thsotoy* was used (consistent with Berlin et al., 1973). On the single remaining occasion, *lelhey* (inhabitants) was used. It is interesting that this was in the context of describing distinctions between the animals of the forest (*thsotoy*) and animals of the water (*tshotoy inot lheley*). It seems plausible that by using *lelhey* in this context, the speaker was able to maintain the distinction between these two habitat-based groups of animals. Using *tshotoy* in this context, to refer to the more inclusive animal category, would likely have made this distinction difficult to maintain.

A covert category of plants? The Wichi's performance in the induction task suggested that they treat PLANT as a cohesive category as well. The adults' justifications provide converging support. Whenever they referred to plants, it was as a single group. For example, one adult justified her distinction between plants and animals in this way: "because all [animals *tshotoy*] can move away and run. If there is something wrong or bad they can run away, in contrast to these [gestures to the plants] that cannot." Others invoked shared spiritual properties (e.g., "because all of them [gestures to plants] have *husek* …"). It is interesting, however, that the Wichi used not a lexical item in these cases but instead an indexical (e.g., "them," "these") accompanied by gestures toward the plants.

Naming the humans. The adults' justifications also converge well with the observation that although there is a term that includes all and only humans (*hin'ulh*), it is rarely used (Taverna et al., 2014). In our sample, adults referred to people four times, and only when a human was the base. The term *hin'ulh* occurred once, *Wichi* occurred twice, and *siwele* (White people) occurred once.

General Discussion

The Wichi's systematic responses in category-based induction, coupled with adults' justifications, reveal a profile of reasoning about the natural world that shares considerable commonalities with that of individuals from other communities but that also reflects the shaping forces of language, experience, and community-held belief systems in reasoning about the natural world. This work, characterized by strong developmental continuities in the organization of folkbiological knowledge, reveals three overarching themes.

First, this work showcases the power of covert categories. This converges well other evidence documenting the inductive power of categories that, although unnamed, nonetheless guide inductive inferences about the natural world (Anggoro et al., 2008; Berlin et al., 1973; Leddon et al., 2008; Waxman, 2005). This work also illustrates for the first time that although the Wichi may name habitat-based groups of animals and plants, the overarching categories that correspond to ANIMALS (HUMAN AND NONHUMAN) and PLANT support inferences of their own.

Second, these results reveal an emerging precision, over developmental time, in how the Wichi reason about living things. For example, weaving together representations to form categories like PLANT and MAMMAL may be best characterized as a protracted developmental process, evident in adults but not yet in the children. Additional work incorporating more items from each of these categories will be required to flesh out this possibility.

Third, this work fortifies the proposal that anthropocentric reasoning is not a universal developmental starting point for reasoning about the natural world. As predicted, Wichi children did not favor humans as an inductive base in reasoning about the natural world. Moreover, the youngest group of children as well as the adults favored nonhuman animals over humans as an inductive base. This outcome, which contrasts with results from urban, Western-educated children and adults (Carey, 1985; Herrmann et al., 2010), converges well with evidence from several other communities, including rural majority-culture and Native American communities in the United States and Central

America (Atran, Medin, & Ross, 2002; Medin, Waxman, Woodring, & Washinawatok, 2010). In an ideal world, it would have been possible for us to examine younger children as well. However, we did not have access to children in this community who had not yet entered school. We speculate that human-centered reasoning would also be absent in younger children. In future work, we hope to test this speculation with 3-year-olds in the Wichi community; this is the youngest age at which the category-based induction task has been used effectively (Herrmann et al., 2010).

Another goal for future investigations will be to examine whether and how the Wichi understanding of the natural world varies as a function of education level. Recall that most of the adult participants in the current investigation had limited Western-oriented education. A goal in our ongoing work is to consider a larger group of adults to pinpoint the influence (if any) of Western education in adults' folkbiological reasoning.

Implications For Science Education

This experimental work, considered in conjunction with related investigations among the Wichi and other indigenous groups, broadens the empirical foundation for theories of conceptual development and for pedagogy. It also has broader implications for early science education.

Decades of research in the developmental and educational sciences indicate that as children enter the formal classroom, they bring with them their existing knowledge and beliefs about the natural environment and that this knowledge, along with the frameworks children use to organize it, should be taken into account in developing curricula (e.g., Anggoro et al., 2008; Astuti et al., 2004; Coley, 1995; Gelman, 1988; Hatano et al., 1993; Herrmann et al., 2010, 2012; Leddon et al., 2008; Rosengren et al., 1991; Waxman et al., 2007, 2014). First, the current results add to the growing body of evidence on children's frameworks for understanding the natural world. The results underscore that educational practices—even in the natural sciences—are not culturally neutral. They also align well with that view that to advance science education it is advantageous to incorporate culturally based epistemological orientations and cultural practices (Bang & Medin, 2010).

Second, this work provides a foundation for building science curricula that not only take into account children's existing knowledge but harness these strengths in the service of learning. This is especially important in cases like Wichi education. Although officially the goal of the Argentine education system is to root the objectives, contents, and methodologies of the Wichi curriculum in the framework of the indigenous, rural context (EIB Document, Cultural and Educational Ministry, Formosa Province, 2006), curricular decisions are often implemented unilaterally by nonnative educators (e.g., Zidarich, 2014).

For example, we have shown here that Wichi children and adults organize their knowledge of the natural world into a system of categories, some based on taxonomic or perceptual similarities, others based on shared habitat, animacy status, and even spiritual status. The results also show the use of these categories in reasoning about the natural world, even if they are not named. These insights can now be used in developing curricula. In doing so, it will be important to bear in mind that organizing animals into habitat-based categories does not preclude an appreciation of taxonomic categories like those featured in traditional Western science curricula. Moreover, we suspect that the Spanish names for these unnamed categories will be readily borrowed as loan words as soon as the Spanish terms (e.g., *animales, seres vivientes*) are introduced in the curriculum. This phenomenon is already evident in the adults' justifications, in which Spanish loan words appear only for categories that are unnamed in Wichi. Finally, we have shown that even the youngest children in the Wichi community have rich knowledge and models for reasoning about the inhabitants of the earth and that these models differ clearly with the anthropocentric views of the natural world more predominant in Western contexts. We endorse any wariness expressed by Wichi teachers and the community at large about adopting into the curriculum predominantly Western models that so often adopt an anthropocentric stance.

Third, the creation of teaching materials that reflect the Wichi's cultural knowledge is essential to a successful curriculum. This is especially true in the context of the Wichi community because these children have very limited access to written media in their native language (books, magazines, games, etc.), perhaps due to the transmission of the Wichi as written language is socially new. We are currently in the process of starting to redress this limitation by designing a series of books about the inhabitants of earth for school-age children. In designing the materials, we are well aware that the images we offer young children shape their representations of the entities in the natural world and the relations among them. An especially clear example comes from research into the effects of children's media in the United States, which has documented (a) that anthropomorphizing nonhuman animals is far more prevalent in the illustrations and text of books authored by members of the U.S. majority culture compared to those authored by Native Americans (Dehghani et al., 2013) and (b) that young children—especially urban children with little direct contact with the natural world—are sensitive to these representations (Herrmann et al., 2010; Waxman et al., 2014). The media we are developing in the context of the Wichi community are consistent with Wichi categorization patterns and their diverse models of reasoning about the natural world.

In conclusion, as we hope to have made apparent, the Wichi are an especially interesting group to collaborate with from both a research and educational perspective. Our work in this community permits us a glimpse into how folkbiological knowledge is shaped in a community in which the native language and a constellation of experiences and belief systems differ from those of the communities studied in most prior empirical work.

We look forward to continuing to work collaboratively with this community to develop a more strongly indigenous curriculum focusing on native species and on culturally valued characters and beliefs.

Acknowledgments

We thank all native teachers from the Wichi Lako School (Laguna Yema, Formosa, Argentina), particularly Aurelia Pérez, for their dedication and commitment to this work. We are also grateful to children and adults who participated in the Wichi Lawet community. Special gratitude goes to Jennifer Woodring for editorial and statistical insights.

Funding

This research was supported by grants from Argentina (PICT (Proyecto de Investigacion Cientifica y Tecnologica, Scientific and Technological Research project) 2254 from the National Agency of Scientific Promotion) to the first author and from the United States (NSF National Science Foundation - BCS (NSF Division of Behavioral and Cognitive Sciences) 0745594 and DRL (Division of Research on Learning in Formal and Informal Settings) 0815020) to the second and third authors.

References

Alvarsson, J. A. (1988). The Mataco of the Gran Chaco: An ethnografic account of change and continuity in Mataco socio-economic organization. *Acta Universitatis Upsaliensis, 11*, 1–314.

Anggoro, F. K., Medin, D. L., & Waxman, S. R. (2010). Language and experience influence children's biological induction. *Journal of Cognition and Culture, 10*, 171–187. doi:10.1163/156853710X497220

Anggoro, F. K., Waxman, S. R., & Medin, D. L. (2008). Naming practices and the acquisition of key biological concepts: Evidence from English and Indonesian. *Psychological Science, 19*(4), 314–319. doi:10.1111/j.1467-9280.2008.02086.x

Arenas, P. (2003). *Etnografía y alimentación entre los toba-nachilamole#ek y Wichi-lhuku'tas del Chaco Central (Argentina)* [*Etnography and feeding among the toba-nachilamole#ek and Wichi-lhuku'tas from the middle of Chaco (Argentina)*]. Pastor Arenas, Buenos Aires: Author.

Astuti, R., Solomon, G. E. A., & Carey, S. (2004). Constraints on conceptual development. *Monographs of the Society for Research in Child Development, 69*(3), vii–135.

Atran, S., & Medin, D. (2008). *The native mind and the cultural construction of nature.* Cambridge, MA: MIT Press.

Atran, S., Medin, D., Lynch, E., Vapnarsky, V., Ucan, E., & Sousa, P. (2001). Folkbiology doesn't come from folkpsychology: Evidence from Yukatek Maya in cross-cultural perspective. *Journal of Cognition and Culture, 1*(1), 3–42. doi:10.1163/156853701300063561

Atran, S., Medin, D. L., & Ross, N. (2002). Thinking about biology: Modular constraints on categorization and reasoning in the every day life of Americans, Maya and scientists. *Mind and Society*, 6(3), 31–64. doi:10.1007/BF02513147

Bang, M., & Medin, D. (2010). Cultural processes in science education: Supporting the navigation of multiple epistemologies. *Science Education 94*(6), 1008–1026.

Bang, M., Medin, D., & Atran, S. (2007). Cultural mosaics and mental models of nature. *Proceedings of the National Academy of Sciences, USA, 104*, 13868–13874. doi:10.1073/pnas.0706627104

Berlin, B. (1992). *Ethnobiological classification: Principles of categorization of plants and animals in traditional societies.* Princeton, NJ: Princeton University Press.

Berlin, B., Breedlove, D. E., & Raven, P. H. (1973). General principles of classification and nomenclature in folk biology. *American Anthropologist, 75*(1), 214–242. doi:10.1525/aa.1973.75.issue-1

Carey, S. (1985). *Conceptual change in childhood.* Cambridge, MA: Bradford Books.

Carey, S. (1995). On the origin of causal understanding. In D. Sperber, D. Premack, & A. J. Premack (Eds.), *Causal cognition: A multidisciplinary debate* (pp. 268–308). New York, NY: Oxford University Press.

Carey, S. (1999). Sources of conceptual change. In E. Scholnick, K. Nelson, S. Gelman, & P. Miller (Eds.), *Conceptual development: Piaget's legacy* (pp. 293–326). Mahwah, NJ: Erlbaum.

Carey, S., Zaitchik, D., & Bascandziev, I. (2015). Theories of development: In dialog with Jean Piaget. *Developmental Review, 38*, 36–54. doi:10.1016/j.dr.2015.07.003

Coley, J. D. (1995). Emerging differentiation of folkbiology and folkpsychology: Attributions of biological and psychological properties to living things. *Child Development, 66*, 1856–1874. doi:10.2307/1131915

Coley, J. D., Medin, D. L., & James, L. (1999, April). *Folk biological induction among Native American children.* Paper presented at the Biennial Meetings of the Society for Research in Child Development, Albuquerque, NM.

Dehghani, M., Bang, M., Medin, D., Marin, A., Leddon, E., & Waxman, S. (2013). Epistemologies in the text of children's books: Native- and non-native-authored books. *International Journal of Science Education, 35*(13), 2133–2151. doi:10.1080/09500693.2013.823675

EIB (Bilingual Intercultural Education) Document, Cultural and Educational Ministry, Formosa Province. (2006). *En enfoque EIB (Educación Intercultural Bilingüe) y la enseñanza de las ciencias de la vida [The EIB perspective and the teaching of Life Sciences].*

Ferry, A. L., Hespos, S. J., & Waxman, S. R. (2010). Categorization in 3- and 4-month-old infants: An advantage of words over tones. *Child Development, 81*(2), 472–479. doi:10.1111/j.1467-8624.2009.01408.x

Fulkerson, A. L., & Waxman, S. R. (2007). Words (but not tones) facilitate object categorization: Evidence from 6-and 12-month-olds. *Cognition, 105*(1), 218–228. doi:10.1016/j.cognition.2006.09.005

Gelman, S. A. (1988). The development of induction within natural kind and artifact categories. *Cognitive Psychology, 20*(1), 65–95. doi:10.1016/0010-0285(88)90025-4

Hatano, G., Siegler, R. S., Richards, D. D., Inagaki, K., Stavy, R., & Wax, N. (1993). The development of biological knowledge: A multi-national study. *Cognitive Development, 8*(1), 47–62. doi:10.1016/0885-2014(93)90004-O

Hays, T. E. (1976). An empirical method for the identification of covert categories in ethnobiology. *American Ethnologist, 3*, 489–507. doi:10.1525/ae.1976.3.3.02a00070

Herrmann, P. A., Medin, D. L., & Waxman, S. R. (2012). When humans become animals: Development of the animal category in early childhood. *Cognition, 122*(1), 74–79. doi:10.1016/j.cognition.2011.08.011

Herrmann, P., Waxman, S. R., & Medin, D. L. (2010). Anthropocentrism is not the first step in children's reasoning about the natural world. *Proceedings of the National Academy of Sciences USA, 107*(22), 9979–9984. doi:10.1073/pnas.1004440107

Inagaki, K. (1990). The effects of raising animals on children's biological knowledge. *British Journal of Developmental Psychology, 8*(2), 119–129.10.1111/bjdp.1990.8.issue-2

Inagaki, K., & Hatano, G. (1993). Young children's understanding of the mind-body distinction. *Child Development, 64*, 1534–1549. doi:10.2307/1131551

Koschitzky, M. (1992). *Las telas de malla de los wichi/mataco: Su elaboración, su función y una posible interpretación de los motivos* [*Wichi/Mataco's mesh fabric: Its development, function and a possible interpretation of the drawings*]. Buenos Aires: Centro Argentino de Etnología Americana.

Leddon, E. M., Waxman, S. R., & Medin, D. L. (2008). Unmasking "alive": Children's appreciation of a concept linking all living things. *Journal of Cognition and Development, 9*(4), 461–473. doi:10.1080/15248370802678463

Leddon, E., Waxman, S. R., Medin, D. L., Bang, M., & Washinawatok, K. (2012). One animal among many? Children's understanding of the relation between humans and nonhuman animals. In G. Hayes & M. Bryant (Eds.), *Psychology of emotions, motivations and actions: Focus on civilizations and cultures series.* Hauppauge, NY: Nova Science.

Ley Argentina de Educación Nacional 26.206. [National Education Law of Argentina]. (2006). *Título II, Capítulo XI: Educación Intercultural y Bilingüe.* Retrieved from http://lengua.com.ar/uploads/media_items/ley-26206-de-educacion-nacional-completa.original.pdf

Maranta, A. A. (1987). Los recursos vegetales alimenticios de la etnia mataco del Chaco centro occidental [Food plant resources of the Mataco ethnic group from middle western Chaco]. *Parodiana, 5*, 161–237.

Medin, D. L., & Bang, M. (2014). *Who's asking? Native science, Western science and science education*. Cambridge, MA: MIT Press.

Medin, D., Waxman, S., Woodring, J., & Washinawatok, K. (2010). Human-centeredness is not a universal feature of young children's reasoning: Culture and experience matter when reasoning about biological entities. *Cognitive Development, 25*(3), 197–207. doi:10.1016/j.cogdev.2010.02.001

Nercesian, V. (2011). *Gramática del Wichi, una lengua chaqueña. Interacción fonología-morfología-sintaxis en el léxico [Wichi grammar of a language from Chaco. Interaction phonology-morphology-syntax in the lexicon]*. (Unpublished doctoral dissertation). Universidad de Buenos Aires.

Palmer, J. H. (2005). *La buena voluntad Wichi: Una espiritualidad indígena [The Wichi Good Will: an Indigenous spirituality]*. Grupo de trabajo Ruta 81, Las Lomitas, Buenos Aires.

Pierotti, R. J. (2011). *Indigenous knowledge, ecology, and evolutionary biology*. New York, NY: Taylor & Francis.

Rosch, E., Mervis, C., Gray, W., Johnson, D., & Boyes-Braem, P. (1976). Basic objects in natural categories. *Cognitive Psychology, 8*, 382–439. doi:10.1016/0010-0285(76)90013-X

Rosengren, K. S., Gelman, S. A., Kalish, C. W., & McCormick, M. (1991). As time goes by: Children's early understanding of growth in animals. *Child Development, 62*, 1302–1320. doi:10.2307/1130808

Ross, N., Medin, D., Coley, J. D., & Atran, S. (2003). Cultural and experiential differences in the development of folkbiological induction. *Cognitive Development, 18*(1), 25–47. doi:10.1016/S0885-2014(02)00142-9

Solomon, G. E. A., & Cassamites, N. L. (1999). On facts and conceptual systems: Young children's integration of their understanding of germs and contagion. *Developmental Psychology, 35*, 113–126. doi:10.1037/0012-1649.35.1.113

Springer, K., Ngyuen, T., & Samaniego, R. (1996). Early understanding of age- and environment-related noxiousness in biological kinds: Evidence for a naive theory. *Cognitive Development, 11*, 65–82. doi:10.1016/S0885-2014(96)90028-3

Stavy, R., & Wax, N. (1989). Children's conceptions of plants as living things. *Human Development, 32*, 88–94. doi:10.1159/000276367

Suárez, M. E., & Montani, R. (2010). Vernacular knowledge of bromeliaceae species among the Wichi people of the Gran Chaco, Argentina. *Journal of Ethnobiology, 30*, 269–292. doi:10.2993/0278-0771-30.2.265

Tarlowski, A. (2006). If it's an animal it has axons: Experience and culture in preschool children's reasoning about animates. *Cognitive Development, 21*(3), 249–265. doi:10.1016/j.cogdev.2006.02.001

Taverna, A. S., Waxman, S. R., Medin, D. L., Moscoloni, N., & Peralta, O. A. (2014). Naming the living things: Linguistic, experiential and cultural factors in Wichi and Spanish speaking children. *Journal of Culture and Cognition, 14*, 213–233. doi:10.1163/15685373-12342122

Taverna, A., Waxman, S., Medin, D., & Peralta, O. (2012). Core folkbiological concepts: New evidence from Wichi children and adults. *Journal of Cognition and Culture, 12*(3–4), 339–358. doi:10.1163/15685373-12342079

Terraza, J. (2009). *Grammaire du Wichi: phonologie et morphosyntaxe [Wichi Grammar: phonology and morphosyntax]* (Unpublished doctoral dissertation). Université du Québec à Montréal, Montreal, Quebec, Canada.

Vidal, A., & Nercesian, V. (2009). Estudio del léxico Wichi (mataco mataguaya): Aportes al conocimiento de algunas situaciones de contacto y desplazamiento lingüístico en el Chaco argentino [*Study of the Wichi lexicon (mataco mataguaya): Contributions to the knowledge of some contact situations and language shift in the Argentinean Chaco*]. *Cuadernos Inter-Culturales, 7*(12), 141–158.

Waxman, S. R. (2005). Why is the concept "living thing" so elusive? Concepts, languages, and the development of folkbiology. In W. Ahn, R. L. Goldstone, B. C. Love, A. B. Markman, & P. Wolff (Eds.), *Categorization inside and outside the laboratory: Essays in honor of Douglas L. Medin* (pp. 49–67). Washington, DC: American Psychological Association.

Waxman, S. R., & Gelman, S. A. (2009). Early word-learning entails reference, not merely associations. *Trends in Cognitive Sciences, 13*(6), 258–263. doi:10.1016/j.tics.2009.03.006

Waxman, S. R., & Hall, D. G. (2004). Everything had a name, and each name gave birth to a new thought: Links between early word-learning and conceptual organization. In D. G. Hall & S. R. Waxman (Eds.), *From many strands: Weaving a lexicon*. (pp. 295–335). Cambridge, MA: MIT Press.

Waxman, S., Herrmann, P., Woodring, J., & Medin, D. (2014). Humans (really) are animals: Picture-book reading influences five-year-old urban children's construal of the relation between humans and non-human animals. *Frontiers in Developmental Psychology, 5*(172), 1–8. doi:10.3389/fpsyg.2014.00172

Waxman, S. R., & Lidz, J. (2006). Early word learning. In D. Kuhn & R. Siegler (Eds.), *Handbook of child psychology* (Vol. 2, pp. 299–335). Hoboken NJ: Wiley.

Waxman, S. R., & Markow, D. B. (1995). Words as invitations to form categories: Evidence from 12- to 13-month-old infants. *Cognitive Psychology, 29*(3), 257–302. doi:10.1006/cogp.1995.1016

Waxman, S., Medin, D., & Ross, N. (2007). Folkbiological reasoning from a cross-cultural developmental perspective: Early essentialist notions are shaped by cultural beliefs. *Developmental Psychology, 43*(2), 294–308. doi:10.1037/0012-1649.43.2.294

Wilbert, J., & Simoneau, K. (1982). *Folk literature of the Mataco Indians.* Los Angeles: University of California at Los Angeles Latin American Center.

Winkler-Rhoades, N., Medin, D., Waxman, S. R., Woodring, J., & Ross, N. O. (2010). Naming the animals that come to mind: Effects of culture and experience on category fluency. *Journal of Cognition and Culture, 10(1)*, 205–220. doi:10.1163/156853710X497248

Woolley, J. D. (1997). Thinking about fantasy: Are children fundamentally different thinkers and believers from adults? *Child Development, 68*(6), 991–1011. doi:10.2307/1132282

Zidarich, M. (2014). *Pareja vulnerable, si las hay: docente originario y docente no originario.* [*Vulnerable Couple, if any: native and non- native teacher*]. Available at http://www.lenguawichi.com.ar/educacion-bilingue/bibliografía].

Zidarich, M. (2006). Chalanero. *Libro de lectura para la Alfabetización inicial [Chalanero. Reading book for Early Literacy].* (2nd ed.). Buenos Aires, Argentina: Universidad de Buenos Aires.

Generation Conservation: Children's Developing Folkbiological and Moral Conceptions of Protecting Endangered Species

Jolina H. Ruckert

ABSTRACT

Research Findings: This study investigated folkbiological concepts that structure children's moral reasoning regarding conservation. Participants ($N = 52$; 7- and 10-year-olds, gender balanced) were interviewed regarding their values, moral obligations, and rights concerns for endangered and extinct animals. Across the 2 ages, children drew on the animal's living status and teleology, as well as ecological relations, to ground their moral reasoning. Developmentally speaking, the younger children employed both nature-centered and human-centered moral reasoning, whereas the older children employed a complex form of moral reasoning that integrated human models of moral thinking with distinct biological understanding. *Practice or Policy*: The findings can improve education and policy that supports the next generation of conservationists. Educators can draw on these findings to develop class material and activities that teach children about ecological relations—knowledge children can employ to construct their moral arguments for conservation and protecting the environment. The findings revealed that children think about and value endangered animals in deep and beautiful ways. Policymakers can draw on the findings to frame their decision making—drawing more deeply on children's values of the natural world to inform policy decisions. Drawing on children's values in the development of environmental policy engages them in the policy process.

Children are the next generation of conservationists. Caregivers, educators, and community members are working to support children's interest in protecting and preserving the environment (Paltved-Kaznelson, 2009; Rideout, Foehr, & Roberts, 2010). There are now several large-scale institutional initiatives that seek to foster children's conservation concerns (e.g., the United States' Great American Outdoors Initiative and Let's Move Outside, Australia's Planet Ark, and China's The Long Green March). To understand and support children's concerns for conservation psychologists and educators need to understand children's environmental and moral conceptions of protecting the natural world (Karpiak & Baril, 2008; Kellert, 2002; Melson, 2001; Myers, 2007). Research has shown that children across cultures and from an early age care about protecting the environment (cf. Melson, 2001; Myers, 2007; Nevers, Gebhard, & Billmann-Mahecha, 1997). Children value nature (Kellert, 2002), are concerned for the welfare of animals (e.g., Myers, 2007) and nonsentient creatures and landscapes (e.g., Kahn, 1992), attempt to coordinate the needs of the natural world with their own (Kahn & Lourenco, 2002; Katcher, 2002; Severson & Kahn, 2010), can both fear and care for a natural entity (Kahn, Saunders, Severson, Myers, & Gill, 2008), and can morally reason about the natural world in increasingly complicated ways over the course of development (Clayton & Myers, 2009). Yet very little is known about the role of folkbiological concepts in shaping and supporting children's moral judgments about protecting the environment. Children's belief systems and understandings about the way the world works deeply shape their moral judgments regarding other people (cf. Wainryb,

1995). How might children's knowledge and beliefs about the biological world structure their moral concern for the environment? Understanding children's developing environmental and moral conceptions can shape the conservation agenda by informing conservation initiatives, social policy, and environmental education. For example, educators could utilize the information as a toolkit for recognizing children's patterns of ecological and biological thought, what frameworks they are working with, and can build from these to support learning and conservation interests.

The current study provides an initial foray into the investigation of children's developing biological understanding that is brought to bear in their moral concerns for the environment. The current investigation involves a quantitative and qualitative analysis of children's moral and ecological reasoning to explore two questions: (a) How do children use biological knowledge to structure their moral reasoning about the environment? and (b) How do children use psychological or human frameworks to structure their environmental moral reasoning? These questions are poised to shed new light on long-standing discourse regarding the biological–psychological distinction in children's folkbiological thinking.

The current study takes a snapshot of folkbiological and moral concepts as they transform from young childhood to middle childhood. To begin to understand the role of folkbiology in moral development the current study looks at this critical period in conceptual development across these two domains. Folkbiological and moral concepts emerge earlier in development, and they begin to take shape in the preschool years. To gain important insights into the nature of conceptual change in preschool thought one can look ahead in the developmental trajectory as an initial step. Studying developmental trajectories allows better characterization and understanding of developmental antecedents (cf. Coley, 2000). To understand the nature and process of conceptual change in which folkbiological meets moral development, the current study takes a first glance at the conceptual development that preschoolers can potentially acquire.

Biological–psychological distinction in folkbiology

Early theory proposed that biological and psychological constructs are not distinct for the young child (Carey, 1985; Piaget, 1932). Piaget (1932) proposed that the young child's (i.e., prior to 7–11 years) thinking was limited to domain-general intellectual immaturity wherein he or she did not possess a theory of biology but instead understood biological processes in terms of psychology. Carey (1985), drawing on Piaget's work, maintained that young children possess a general domain of understanding in which biological understanding is not distinct from psychological knowledge. She argued that children use their commonsense understanding about psychology to explain and predict biological phenomena until the age of 10. Unlike Piaget, Carey believed that this was due to a lack of knowledge specific to the biological domain (not a domain-general intellectual immaturity as espoused by Piaget). Carey assumed that children, having the most experience and familiarity with humans, would utilize their understanding of humans as the base reference point with which to understand other biological entities (what Carey referred to as the comparison-to-people model). Findings have shown conditional support for Carey's comparison-to-people model, insofar as it has been replicated in urban populations (e.g., Atran, 1998; Coley, 2000; Inagaki, 1990; Medin & Atran, 2004; Ross, Medin, Coley, & Atran, 2003; Waxman & Medin, 2007). That is, the comparison-to-people model is not a universal developmental trajectory. Urban children tend to utilize humans as a preferred reference point (using the human to make generalizations about other animals), whereas children in rural (Waxman & Medin, 2007) and indigenous (Ross et al., 2003) communities have shown a different pattern of generalization (drawing on ecological knowledge). In middle childhood (6–10 years old) urban children begin to move away from the preference for human-centric generalizations (Carey, 1985; Ross et al., 2003; Waxman & Medin, 2007).

Contrary to Carey's (1985) and Piaget's (1932) early claims, it appears that children from an early age have a distinct biological framework. Preschoolers can understand bodily traits as distinct from mental traits (Springer & Keil, 1991). They understand that mental effort is needed to influence mental outcomes such as memory but that this is irrelevant for biological outcomes such as digestion (Inagaki & Hatano,

1993). Children as young as 3 years of age understand that animals are living things (Huang & Lee, 1945) and will attribute to animals biological characteristics, including breathing, growing, reproducing (Backscheider, Shatz, & Gelman, 1993; Greif, Nelson, Keil, & Gutierrez, 2006), and having emotions (Gelman & Spelke, 1981). Four-year-olds utilize vitalistic explanations (i.e., "the notion that living things take vital energy from the environment and use it to grow and function properly," Hatano & Inagaki, 1994; e.g., organs maintain biological activity) for biological phenomena (e.g., breathing; Coley, 2000, p. 84). Five-year-olds also utilize vitalistic explanations. They can reason that living kinds (e.g., squirrel, pine tree, tulip) need to take in food and water to survive (Inagaki & Hatano, 1996; see also Hickling & Gelman, 1995). Eight-year-olds have clearly defined divergent attribution patterns for psychological (e.g., feels happy) and biological (e.g., has bones) properties for the same animals (including predatory animals, such as tigers, and domestic animals, such as turtles; Coley, 1995).

It appears that across cultures and from an early age children possess distinct biological and psychological constructs that they utilize to make sense of the world. The current study seeks to offer the field of folkbiology new means of exploring the interplay of biological and psychological constructs in children's thinking—the ways in which these concepts emerge in distinct forms and in concert in children's environmental moral reasoning. The next section discusses a core distinction observed in children's environmental moral reasoning that can shed new light on understanding the biological–psychological distinction in folkbiology.

Anthropocentric–biocentric distinction in environmental moral reasoning

Research on environmental moral reasoning has revealed children's use of two overarching forms of moral reasoning: anthropocentric and biocentric (Clayton & Myers, 2009; Kahn, 1999; Kahn & Lourenço, 2002; Kahn, Severson, & Ruckert, 2009; Nevers et al., 1997; Severson & Kahn, 2010). Anthropocentric reasoning refers to an appeal that the natural world has moral standing due to the import the natural world has in human lives. For example, one child gave the following justification for why it is wrong to pollute the local waterway: "[It's not alright] because some people that don't have homes, they go and drink out of the rivers and stuff and they could die because they get all of that dirt and stuff inside of their bodies" (Kahn, 1999, p. 100). Biocentric reasoning refers to an appeal that the natural world has moral standing independent of its value to humans. For example, one child reasoned that it is wrong to pollute the water because "fishes, they want to live freely, just like we live freely ... They have to live in freedom, because they don't like living in an environment where there is so much pollution that they die every day" (p. 101). Within biocentrism a form of thinking—*anthromorphic biocentrism*—emerges wherein moral concerns that are directed to the animal are developed through human comparisons ("it's not all right because they're dead, living things just like we are, you wouldn't want anybody to kill you like that," Kahn, 1999, p. 1093). Where anthropocentrism involves moral concerns about humans, anthromorphic biocentrism is moral concern about animals that makes comparisons to humans to ground the animal-centric moral regard. The current study employs a quantitative and qualitative analysis to explore age group differences in anthropocentrism and anthromorphic biocentrism. This analysis seeks to offer the field of early education an account of developing environmental constructs that guide and build children's moral concerns for biodiversity conservation. Discussion on the findings from the current study will offer suggestions for how an understanding of children's folkbiological and environmental moral development can structure environmental education and conservation policies and initiatives.

Methods

Participants

The participants were 52 children: 26 seven-year-olds (*M* age = 7.2 years, *SD* = 0.24; 13 females and 13 males) and 26 ten-year-olds (*M* age = 10.5 years, *SD* = 0.26; 13 females and 13 males). Participants were recruited through a subject pool administered by a university in the U.S. Pacific Northwest and received $10 for their participation. The sample was primarily White (74%), with the remaining children representing a variety of ethnic groups (4% Asian, 2% Hispanic, and 12% mixed descent). Most (56%) children had a reported parental income level of more than $100,000, 32% had a parental income of between $50,000 and $100,000, and 4% had a parental income of $20,000 to $50,000. The median income in 2010 for the county from which participants came was $65,383 (Office of Financial Management, 2013). Moreover, 42% of the children were reported to have at least one parent who had a graduate degree as the highest degree achieved; 46% of parents had an undergraduate degree, 4% had a junior college degree, and 2% had a high school degree. Most parents (83%) reported that their child lived in a two-parent/guardian household. Most parents (92%) reported that the child had siblings (N = 49, X = 2.55, SD = 1.10, range = 0–6). Many (77%) had at least one pet in their household (M = 2.56 pets, SD = 3.48, range = 0–14), and 42% reported having a dog. Children spent time outside on a school day (N = 49, X = 1.79 hr, SD = 0.62, range = 1–3) and on weekends/holidays (N = 49, X = 3.71 hr, SD = 2.0, range = 1–10). Parents provided written informed consent and children provided written assent to all procedures.

The current study focused on the transition from young childhood to middle childhood (between 7 to 10 years old), as it is an important one in the domains of moral and folkbiological conceptual development. At the end of young childhood early forms of universal fairness claims take shape as children move away from egocentric fairness concerns (Damon, 1977; Nucci, 2001). In folkbiology this developmental period marks the move away from human-centric thinking toward a biological orientation (Ross et al., 2003; Waxman & Medin, 2007).

Procedure

The methods presented here were drawn from a larger investigation of children's environmental moral development. Only procedures relevant to the current investigation are presented. Each participant was individually interviewed. The procedure comprised a semistructured interview lasting approximately 35 min. The interview focused on children's moral reasoning in the context of humans harming an animal species. The harm was presented within the context of the ultimate harm to an animal species—the threat of extinction. Children as young as 4 can understand extinction in terms of a combination of death concepts, including that the species no longer persists (Poling & Evans, 2004), and there is indication that animal extinction is conceived of as a moral concern for young children (Grace & Ratcliffe, 2002; Kahn, 1999).

The current investigation focused on data from 10 interview questions in which children's reasoning was elicited. The 10 questions covered three core moral constructs: values, obligations, and rights. First two value-oriented questions were asked (i.e., "Does it matter to you if an animal becomes endangered?" "Does it matter to you if an animal becomes extinct?"). Following this children's moral obligatory judgments regarding humans harming the endangered gray wolf were assessed using three criterion judgment questions. The gray wolf was chosen as it was a familiar and charismatic endangered species (Kellert, 1996) that was found in the Pacific Northwest. The assessment of moral obligation drew on domain theory research on moral obligatory judgments that are prescriptive, not contingent on rules or authority; are generalizable across relevant contexts; and are based on justifications of welfare, fairness, and rights (Helwig, 1995; Smetana, 1981; Turiel, 1983; Turiel & Davidson, 1986). Thus, the first question focused on whether killing the gray wolves was viewed prescriptively (i.e., "Is it alright or not alright for people here in Washington State to kill the gray wolves?"). The second question focused on the rule contingency of the judgment (i.e., "Let's say that there was a law here in Washington State that allows people to kill the gray wolf. With this law in place, would it then be alright or not alright for people here in Washington State to kill the gray wolves?"). The third question focused on whether the judgment regarding killing the gray wolf generalized across contexts where a society legitimated the practice (i.e., "Let's say that far away in the country of Greenland people kill gray wolves. Would it then be it alright or not alright for people in Greenland to kill the gray wolves?").

The final five questions (developed from Kahn et al.'s, 2008, methodology) focused on children's conceptions of animal rights (developed from Kahn et al.'s, 2008, methodology). Participants were asked about their understanding of animal rights (e.g., "Do you think animals have rights or do not have rights?" "Do you think that gray wolves have rights or do not have rights? [if the child stated "have rights": "Which ones?"]). Additional rights-based questions focused on specific rights, including the right to live (i.e., "Do you think that gray wolves have the right to not be killed?") and the right to freedom (e.g., "Do gray wolves have the right to be wild?" "Should the gray wolf be allowed to be wild and live free?").

Coding and reliability

The construction of the coding manual followed well-established methods in the social-cognitive literature (Damon, 1977; Kohlberg, 1984; Turiel, 1983) and largely the previous coding manuals developed on children's environmental moral justifications (see Kahn, 1999, for a review). The entire set of interviews was coded. A second researcher then independently recoded 30% of the data (balanced across age and gender). Intercoder reliability was assessed using Cohen's kappa, a measure of interrater agreement, and was found to be .64 ($p < .001$). A commonly referenced benchmark for interpreting Cohen's kappa is Fleiss, Levin, and Paik (2003), who rated any value of kappa over .75 as excellent agreement, between .40 and .75 as intermediate to good, and below .40 as poor. Previous research has reported kappas for reasoning data between .63 and .78 (see Kahn, 1999; Severson & Kahn, 2010). See Table 1 for abbreviated descriptions of the justification categories.

Coding is the systematic approach to interpreting and characterizing justification data. Justifications are the varied reasons children give for the judgments they make. Multiple codes were possible for justifications. For example, a child could provide in one response a statement regarding both anthropocentric and anthromorphic reasoning. In such a situation both codes would be marked. All significant justifications were coded instead of only the most predominant justification. A given justification was only coded once for each question. A justification that a child used on the basis of a misunderstanding of a question was not coded. An uncodable category existed for justifications that were too unintelligible or incomplete to code, responses in which a participant answered a question other than the one that was asked, cases when forcing the response into an existing category would have distorted the meaning of the response, and cases when a participant gave an "I don't know" justification.

Coding justification data is a challenging task and requires more nuanced interpretations than coding other forms of interview data (e.g., evaluations or content coding). In interpreting a response it may be required to read beyond the particular question and review the preceding and following questions or the overall tenor of the responses that arises over the entirety of the interview. In such cases when an unclear justification is clarified later in the interview, it is permissible to recode

Table 1. Justification Categories.

Category	Definitions and Examples From Data
Anthropocentric	An appeal to the welfare of human beings ("animals can be important, um, like, in the human, um, like, the human cycle. So, like what we eat and stuff.")
Biocentric	
Ecocentric	An appeal for the welfare of biological life and ecological systems ("they don't want to be killed ... they don't want to die," "because they're wild animals and that's like in their blood," "because it's some animal's food," "when they die, um, helps some of the plants grow, their poop does it too")
Anthromorphic	An appeal that is based on recognizing a correspondence between the moral standing (including claims to respect, fairness, rights, or intrinsic value) of humans and that of other natural biological or nonbiological entities. Included here are reasons wherein an inequivalent correspondence between nature and humans does not void the mapping of similar biocentric moral considerations. ("they are living things too like us, ... they can think too and they know when they are being hurt," "'Cause animals are creatures just like we are and they have the same rights as we do. [*How come?*] Well, not necessarily the same rights, um, to vote, but they have the same rights to live on planet earth").

justifications, but this rule must be used judiciously and conservatively (see also Kahn, 1999; Kahn & Lourenço, 2002) and has the potential to challenge intercoder reliability scores.

Results

Most children endorsed proenvironmental judgments across the 10 questions that garnered reasoning data. That is, children stated that it mattered if an animal became endangered (96%) or extinct (98%); that it was not alright to kill the gray wolf (90%), even if a law allowed it (94%) or even if it was a cultural practice (83%); that the gray wolf should be allowed to be free (90%); that animals have rights (94%); that gray wolves have rights (94%); and that gray wolves have the rights to not be killed (86%) and to be wild (87%). Children's justifications were elicited for these 10 questions. The resulting justifications were coded with the categories reported in Table 1. The quantitative results are reported in Table 2, broken down by each of the 10 questions and the age groups.

The results of the justification questions were then used to assess the degree of (a) anthropocentrism, (b) biocentric ecocentrism, and (c) biocentric anthromorphic reasoning endorsed across the 10 questions. To provide an overall assessment of the degree to which children used the three justification categories, and to test for developmental differences, I summed the 10 justification questions to three scores reflecting the number of times the children used each of the three categories across the 10 questions. For anthropocentric (human-centered) reasoning, the 7-year-olds had a mean of 4.38 ($N = 26$, $SD = 2.22$, $SE = .44$) and the 10-year-olds had a mean of 2.15 ($N = 26$, $SD = 2.05$, $SE = .40$). For biocentric ecocentrism, the 7-year-olds had a mean of 3.15 ($N = 26$, $SD = 2.84$, $SE = .55$) and the 10-year-olds had a mean of 4.81 ($N = 26$, $SD = 2.6$, $SE = .52$). For biocentric anthromorphism, the 7-year-olds had a mean of 0.73 ($N = 26$, $SD = 1.31$, $SE = .26$) and the 10-year-olds had a mean of 2.38 ($N = 26$, $SD = 2.71$, $SE = .53$). An independent-samples t test revealed that 7-year-olds endorsed anthropocentric reasoning to a greater degree than 10-year-olds, $t(50) = 3.75$, $p < .001$. An independent-samples t test revealed that 10-year-olds endorsed ecocentric reasoning to a greater degree than 7-year-olds, $t(50) = -2.18$, $p = .03$. An independent-samples t test revealed that 10-year-olds endorsed anthromorphic reasoning to a greater degree than 7-year-olds, $t(50) = -2.80$, $p = .007$.

Quantitative analysis of biocentric moral reasoning

Comparing anthropocentrism and biocentrism

As an additional step, a 2 (age) × 2 (reasoning category) repeated measures analysis of variance was then conducted to look at differences in the use of anthropocentric and biocentric ecocentrism. This comparison looked at the children's use of human-centered moral reasoning and nature-centered moral reasoning. That is, it compared the use of human welfare concerns to nature welfare concerns. There was no main effect for age, $F(1, 50) = 0.59$, $p = .45$. There was a main effect for reasoning type,

Table 2. Percentages of Proenvironmental Justification by Category.

		Value of Nature		Moral Obligatory Criteria				Conception of Animal Rights			
Justification Category	Age	Matters if Animals Become Endangered	Matters if Animals Become Extinct	Not Okay to Kill	Not Okay to Kill With Law	Not Okay to Kill in a Faraway Country	Gray Wolf Should Be Wild and Live Free	Animals Have Rights	Gray Wolves Have Rights	Wolves Have the Right to Not Be Killed	Wolves Have the Right to Be Wild
Anthropocentric	7	73	81	58	58	42	27	12	23	39	27
	10	42	50	35	27	19	19	8	4	12	
Ecocentric	7	19	8	19	23	35	58	35	39	31	50
	10	37	37	52	41	41	70	56	44	44	67
Anthromorphic	7	4		4		4	8	15	15	15	8
	10	12	16	27	15	23	27	35	27	31	23

Note. Percentages do not always equal 100% as multiple justifications were possible.

$F(1, 50) = 13.89$, $p < .001$. There was an Age × Reasoning interaction, $F(1, 50) = 17.88$, $p < .001$. There was a significant difference in that 7-year-olds predominantly used anthropocentric (human-centered) moral concerns, and 10-year-olds predominantly used biocentric ecocentrism (nature centered).

Comparing ecocentric biocentrism and anthromorphic biocentrism

As an additional step, a 2 (age) × 2 (reasoning category) repeated measures analysis of variance was then conducted to look at differences in the use of ecocentrism and anthromorphism. This comparison looked at the children's use of nature-centered moral reasoning that referenced a human framework and nature-centered moral concerns that did not. There was a main effect for age, $F(1, 50) = 10.30$, $p = .002$. Ten-year-olds used biocentric reasoning more than 7-year-olds. There was a main effect for reasoning type, $F(1, 50) = 29.28$, $p < .001$. Both 7- and 10-year-olds predominantly used ecocentrism. There was no Age × Reasoning interaction. The degree to which 7-year-olds used ecocentrism more than anthromorphism was not significantly different from the degree to which 10-year-olds used ecocentrism more than anthromorphism ($p = 1.00$).

Qualitative analysis of biocentric moral reasoning

The following qualitative presentation of justification data highlights types of folkbiological understanding that children brought to bear in their environmental moral reasoning. The two overarching forms of biocentric reasoning—that is, ecocentric (concern for the welfare of biological life and ecological systems) and anthromorphism (concern for nature that is based on human referential reasoning)—are addressed here. The qualitative analysis first addresses the distinct biological knowledge that emerged in children's ecocentric reasoning, followed by an analysis of the biocentric (nature-centered) reasoning that drew on human frameworks to ground moral standing for nature.

Distinct biological knowledge in ecocentric moral reasoning

Biological life and living status

The animal's living status was referenced to support moral concerns. For example, a 10-year-old female stated, "[Animals] should have rights … because they're living things." Similarly, a 10-year-old male stated, "I think they do have rights. [*Why?*] Um, because they are living things, and all living things should have rights." A 7-year-old boy stated that gray wolves have the right to not be killed "'cus every animal deserves to be on the world. [*How come?*] Because they're cool, they, they have a life." These children grounded moral concern on the living status of the animal.

Ecological construals

Children made appeals to the inherent worth of natural processes, such as the food chain or other natural balances in an ecosystem. A 10-year-old girl stated that it matters if animals become endangered: "Because that animal might be really good for the environment, like it might eat things like pesticides, for that, so that it can help protect other animals. So it might eat bad bacteria for other animals too." A 7-year-old boy stated, "[It matters to me if an animal becomes endangered] 'cause some of the animals help the earth. [*How so?*] Some of them fertilize the plants and when they die, um, helps some of the plants grow, their poop does it too." A 10-year-old boy stated, "Like if trees, all pines went extinct, then that would be a lot of the oxygen-producing things in the world." A 7-year-old boy stated, "[Gray wolves should be wild and free] because it's some animal's food." A 10-year-old female described the food chain in depth:

> Well, if there was a law, I guess it'd be okay, but it's still not alright [to kill the gray wolf], but I still wouldn't do it. [*Okay, why not?*] 'Cause they're a part of, they're part of the food chain, so. [*Okay, can you tell me more about that?*] The zebra eats grass, the lion eats a zebra, the lion, the turkey vultures eat lions, turkey vultures, I don't know, and they turkey vultures die and they go back to the soil and the soil just goes into the grass and then the.… then they turn and they come back down so there's a pile of them and it goes into the soil grows grass and then it starts all over

again. [*Nice.*] And like the the worms are in the soils and the birds eat the worms and then like turkey vultures eat birds. [*So is the food chain important?*] Yes. [*How come?*] Uh, let's say there wasn't anymore grass, so then they um, I don't know what I used after grass, I used the zebras and so then the zebras don't have anything to eat and so then all the zebras would die and then 'cause the lions don't have any more zebras to eat and the lions would die and 'cause the turkey vultures didn't have any lions to eat, the turkey vultures would die.

Along similar lines, a 10-year-old boy stated the following:

> [It matters if an animal becomes endangered] because it um, it'll affect the um, whole world, because of the food chains. [*Okay.*] And- [*Tell me about that.*] It's when they, um, it all starts with the sun. And then like, it goes to the grass, where the photosynthesis and then, if the deer become extinct, they'll be no, um, herbivores to eat the grass and the, um, animals that prey on the deer, well, probably won't have that much food left. [MMMM HMMM] [*Okay.*]. [*So why does that matter?*] Because then more and more animals will die and become endangered and extinct.

Teleological claims

Children supported moral judgments with teleological reasoning, or appeals to the natural entities' design or purpose. For example, a 10-year-old female stated that gray wolves should be wild and free "because that's where they're supposed to be wild things, I guess, and I think they should be able to live there." For this young girl, the wild status of the animal grounded a teleological argument that the animal is meant to be in the wild and that is where "they should be." A 10-year-old boy stated that gray wolves have the right to be wild "because they were wild to begin with." This boy made an appeal to the original state of the animal to justify its continuation. This is a form of is-to-ought reasoning wherein the way that something has been necessitates how it should be. Another 10-year-old boy stated, "Because they're wild animals and that's like in their blood." The child employed an English colloquialism that implies a teleology based on biology or inheritance. One 10-year-old boy stated, "Because they are wild animals and not like domesticated." The boy labeled the wolf as wild, not domesticated, and placed the wild animal, not the domesticated animal, in the wild environment. Finally, consider this 10-year-old female's teleological account of wildness:

> I don't think it should be kept in a backyard, but ah, um, and it shouldn't be fenced off, because this is nature, and this is what nature does, I mean, I don't think that the wol- 'cause if they are put in fenced off areas then they can't they they have if they have really big territories then they may like, in the zoo, like animals pace, and that usually means 'cause that they're bored and they don't know what to do. And that would be a bit like an animal in captivity, then it's not itself, it's just something that it's made to be ...

This child also suggested that wild animals are meant to be in the wild. A wild animal that is in captivity is not fulfilling the teleological purpose of the wild design: It is "not itself, it's just something that it's made to be." When a wild animal is not in the wild, perhaps fenced off, it does not have access to the large territories that wild lands provide. Wildness is tied to a biological account of flourishing for the wild animal. A teleological account recognizes that fulfilling one's design or purpose engages the individual's potential to flourish.

Justice-oriented reasoning emerged contextualized within a teleological account, wherein the child's moral concern established fundamental linkages between justice concepts and teleology such that the entity has the right to fulfill a purpose or design. Consider a 10-year-old female's judgment regarding the rights of gray wolves:

> ... they can't vote, but they can have their natural rights. [*And why do they have their natural rights?*] 'Cause that's just how the world works and that's just how they're gonna not become extinct. [*Tell me more about what you mean about "how the world works."*] Well, that's 'cause, if you're um, it's how, humans, how their way is is ta come up with things and do the things that we do, but that they're still um, like the way the world is to do that, but the way of the world for wolves is also to do what they do, so that's kind of what I mean by the way of the world. [*Which rights does a gray wolf have?*] Their territories, and their mating, and um, and um, maybe what they eat, because when they, something that they would eat, would be, the elk and the stuff, and their elk and sometimes berries and stuff, like that, but um, also if that's like the um, if that's the way, how, it it, goes, then it's the way of the food chain, and their few rights come together because that's how it goes, and so they have their right to mate, because that's their way of reproducing.

The child suggested that the wolves have "natural rights" that are based on the design or purpose of the wolf. For example, the wolves would have the right to mate, as that is their way of reproducing. The wolves have the right to eat, as that is the way of the natural process of the food chain, in which the wolves eat elk and elk eat berries. To fulfill the design of their link in the food chain, they have the right to eat. The rights granted to the wolf are granted based on the "way of the world" and the "world of the wolves" insofar as these worlds define a teleological account of biological design.

Anthromorphic biocentric reasoning

Children would ground their moral concern for the animal in reasoning that built on comparisons with humans. Consider this 10-year-old male's justification for why it is wrong to kill the gray wolf: "They are living things too like us, ... they can think too and they know when they are being hurt." This child grounded the concern for the intrinsic value of the life of the wolf on the capacity to experience pain, an experience shared by humans. Similarly, a 10-year-old female stated that it would matter if an animal became extinct "because those animals, they have feelings and they are like us, and they have feelings ..." Along the same lines, this 10-year-old female stated, "Because well it's basically like killing anything else or anyone else and that's not allowed ... because they're living things ... because they're just like you and you wouldn't want people coming around and killing you." The children were taking the perspective of the animal's experience as a subjective being with the capacity to feel in ways similar to their own experiences, such as also having the capacity to "think" and "know when they are being hurt."

Children's conceptions of human rights were employed to ground their conception of animal rights. For example one 10-year-old male stated that wolves have the right to be free "because they don't like to be pinned up or killed. They like to be free like us." Just as humans prefer certain freedoms, wolves too prefer certain freedoms. A 10-year-old female stated

> I think animals have rights because it's important for animals to be free, like we're a free country and animals should be free too. [*How come?*] 'Cause then that's kinda fair, like we're free, we still have laws, but we're free.

For this young girl granting humans freedom warranted granting freedom to other animals, within the jurisdiction of a "free country." One 7-year-old male engaged in a justice-oriented isomorphism, stating, "[Animals have rights] because humans are animals, they should be treated equal, like humans, we are animals too, we should, all animals should be treated equal ..." Here the child drew on the concept of equality, extended equality to all animals, and based the extension of equality concerns on what is granted to humans.

At times the comparisons between humans and the natural world were based on both similarities and differences, yet the qualities that establish similarity were overriding in establishing moral standing of the animal. For example, consider this 10-year-old female's justification for why it is wrong to kill the gray wolf:

> They do have a heart and they do have, I mean they do have things to do in their life and although they may not make checklists of what they have to do in that day, they also know well, we need to go, 'cause that's their instinct, they have to go and find food and in the springtime find a mate and everything ...

The young girl recognized that the animal has needs to fulfill ("they do have things to do in their life") that may be different from human needs ("they may not make checklists") but are legitimate and deserving of regard. A 10-year-old male's transmorphic justification paralleled the young female's concerns for the animal's experiences that are at once similar to and different from humans'. This child found meaning and power in the coupling of similarities and differences:

> I think that animals are smart in a sort of way, like in a different kind of way, than we are. Like we are smart in communication, they are smart in communication, they like use howls, but we like use words. And we have a more complicated language. And sometimes we think that we are smarter than animals, but they might have a type of smartness that we do not have, and stuff like that. And they- like a lot of animals are really good at smelling. We're not that good at smelling. But like, if I wake up and I start going down the stairs and I start

> smelling, I can smell, like if I'm going down the stairs and like the kitchen is like three rooms away, I can smell the smell of pancakes or stuff like that. But I bet wolves and dogs definitely smell better.

The young boy recognized that there are differences between humans and wolves: In some cases wolves have an advantage over humans ("animals are really good at smelling. We're not that good at smelling"), and in some cases the human capacity exceeds the nonhuman capacity ("we have a more complicated language"). The differences are based on similarities, like both animals communicate, have "smartness," or can smell.

One 10-year-old male stated, "'Cause animals are creatures just like we are and they have the same rights as we do. [*How come?*] Well, not necessarily the same rights, um, to vote, but they have the same rights to live on planet earth." Here the concern for equity of rights emerged. They may not be the same rights in all cases, but both humans and animals possess the right to live. A 10-year-old girl captured this idea when she granted that a wolf has the right to be wild:

> Yes. Well it depends what wild means, if wild means, you know, crazy, well then they don't really have that right. But you know if it's to be wild and what they normally do, then I think they do, 'cause what I normally do is maybe sit down in the summer read a book outside and I have my right to be to draw or my mom has her right to go and if she wants to go and start up her own business, she has her right to do that, and that would mean that the wolves would have their right to go and do the things that they do.

Just as humans have rights to do certain things (e.g., draw or own a business), wolves too have rights to do the things that wolves do, such as "to be wild."

Discussion

Through a combination of quantitative and qualitative analyses I can begin to paint a picture of the sorts of folkbiological constructs that structure children's developing moral reasoning for endangered species. Children's moral reasoning emerged in three overarching constructs: (a) distinct human construals, (b) distinct ecological construals, and (c) construals that compared and contrasted human and ecological constructs. Quantitative results reveal age group differences in these forms of reasoning. It appears that with age children use human-centered reasoning less and more nature-centered reasoning that employs human frameworks to construct the concerns. Children's moral concerns in early middle childhood are not about human welfare—they are about animal and ecological welfare—but the moral regard is increasingly structured with referential comparisons and contrasts to human experience. Whereas the young children are grounding their moral regard in distinct human construals or distinct ecological construals, the onset of middle childhood sees distinct human construals give way to a combinatory form of reasoning that builds on human and ecological frameworks. The reasoning is not personification, in which young children attempt to understand behaviors of plants and animals by attributing human processes and properties, an intellectual immaturity posited by Piaget (1932). It is not a matter of a lack of domain-specific knowledge in biology, as posited by Carey (1985). Rather, it is an advanced form of reasoning that integrates conceptually the discrete constructs. For example, some rights are based on biological needs. Animals have biological needs. The animal's biological needs may be similar to or different from humans', but the inequivalent correspondence between nature and humans does not void the mapping of similar moral considerations.

The findings support the argument that conceptual development involves discrete domains that form early in children's thinking and become elaborated and increasingly integrated with development. Keil (1986) posited that folkbiological concepts are differentiated early in development and are replaced by increasingly complex hierarchically integrated conceptual concepts in later development (see Vosniadou & Brewer, 1992, for findings on earth concepts across age groups and an integration of discrete concepts observed in younger children compared to integration of Earth concepts in older children). In the current study younger children tended to state that nature has moral standing in terms of its import to humans (e.g., "animals can be important, um, like, in the ... like what we eat and stuff"), whereas older children argued for the welfare of the animals, not humans, and this reasoning was constructed with parallels to arguments for human moral concerns. It appears that the human framework for moral thinking acts as a toolkit for reconstructing moral ideas away from human-centered concerns toward nature-centered concerns. Consider, for example, the ways in which children based their concerns for the animal's sentiency on their understanding of the human capacity to feel pain. The children likened the animal's experiences to human experiences of feeling pain ("you wouldn't want people coming around and killing you," "they are living things too like us, ... they can think too and they know when they are being hurt"). The children were taking the perspective of the animal's experience as a subjective being with the capacity to feel pain in ways similar to their own experiences. Taking the perspective of an animal can garner nature-centered moral concerns in young adults (Berenguer, 2010) and children (Almeida, Vasconcelos, Strecht-Ribeiro, & Torres, 2013).

Along similar lines, biological needs of the animal were at times construed within the context of human needs. Recall the young girl who recognized that the wolves have needs to fulfill ("they do have things to do in their life") that may be different from human needs ("they may not make checklists") but are legitimate and deserve consideration. Or recall the child who recognized wolves' capacities for communication and intelligence ("they are smart in communication, they like use howls") and drew on his own capacities for communication and intelligence ("... but we like use words. And we have a more complicated language"). The older children found both similar and dissimilar features and then subsequently grounded their moral regard for the nonhuman animal in the complementary nature of similarities and meaningful inequalities between humans and nonhuman animals.

The interplay between the 10-year-old children's distinct biological knowledge and their human frameworks emerged in their rights concerns. For example, recall the 10-year-old who referred to the wolves' "natural rights," such as the right to mate and to eat in order to maintain the rhythms of the species, its reproduction and viable populations, and for balance with greater systems, namely, the food chain. The concern for specific needs and behaviors is contextualized by an account of fulfillment at a species level and in terms of the species in relation to the wider ecosystem. The wolves' rights are not like the human right to vote but are specific to the wolf and its capacity to achieve its biological design and placement in the wider system. The wolves have a right to be wolves. A human has the right to be a human and fulfill the capacities that make him or her human. Similarly, recall the 10-year-old who stated that just as she has the right to draw, or her mom has the right to own a business, the wolf has a right to be a wolf and do what wolves normally do. The older children's anthromorphic biocentric reasoning integrates distinct biological constructs, such as animal needs and capabilities, with the concept of moral arguments based on human needs and capabilities, into a nuanced and complex conception of moral concern extended to the nonhuman animal. It appears that with age, the human framework is conceptually hierarchically integrated with distinct biological knowledge to ground moral concerns for animals.

The qualitative findings reveal, for the first time in the literature, the discrete folkbiological concepts that structure moral reasoning. Findings reveal, unlike Carey's (1985) and Piaget's (1932) early claims, and consistent with more recent findings in folkbiology, that young children, prior to the age of 10, have distinct knowledge about the biological world. Key biological presuppositions that emerged in children's moral reasoning included concepts around (a) living entities that possess biological needs and capabilities, (b) ecological relations, and (c) teleological explanations regarding essence and design. These ecological constructs were used independent of human constructs and were used to ground moral concern for protecting endangered species.

Unlike Piaget (1932) and Carey (1985), who underestimated children's biological knowledge, Keil (1995) posited that young children have a discrete biological understanding and are sensitive to biological needs and abilities. Children in the current study possessed a rich biological knowledge of the endangered species they were interviewed about. The children had knowledge about animal needs and abilities and how these may impact welfare and thus necessitate concern and regard. If an animal needs space to roam, then the animal should have the right to live free and be wild, for example. The wild animal that does not get the room to move, such as the animal in the zoo, does not fulfill its biological needs and may suffer.

Vitalism is the notion that living things require vital energy from the environment to grow and function properly (Hatano & Inagaki, 1994), such as a seed needing rain and sunshine to grow (Hickling & Gelman, 1995) or a squirrel needing food and water to survive (Inagaki & Hatano, 1996). Children in the current study would employ vitalistic explanations in their moral reasoning, particularly with regard to the need to take in vital energy from oxygen and food ("they have to go and find food"). Often references to food needs were made within an ecological explanation. Moral regard was extended to one animal based on food web relations that affected many animals. These urban children rarely made reference to humans as being part of the larger ecological relations. Wider ecological relations that include humans within an ecological system tend to emerge in indigenous and rural, rather than urban, populations (e.g., Coley, 2000; Medin & Atran, 2004; Ross et al., 2003).

Teleological explanations involve assumptions about objects and their parts possessing purpose (Kelemen, 1999). Teleological reasoning that comprises biologically specific assumptions about the physiological nature or self-serving nature of biological functioning is on board as early as preschool (Keil, 1995; Springer & Keil, 1991). Both younger and older children in the current study were using nature-centered moral reasoning that was constructed with concepts of teleology. Children in the current study possessed assumptions around what wolves were meant to experience, such as freedom to roam or reproduce, and these teleological explanations structured rights claims.

Implications for preschool cognitive development

Although a future direction for this research would entail investigating folkbiological constructs in preschoolers' early moral ideas, the current findings can be employed now to support environmental education and social policy that supports young children, including preschoolers. Preschoolers are developing and experiencing meaningful social and moral relationships with animals. Preschool children dream of animals, such as frogs, birds, cows, and sheep (Resnick, Stickgold, Rittenhouse, & Hobson, 1994). Moreover, 20% of young children's invisible imaginary companions are animals (Taylor & Carlson, 1997). Preschoolers judge it wrong to inflict harm on an animal (Zelazo, Helwig, & Lau, 1996). Myers (2007) argued, building from observational studies conducted in a classroom, that preschoolers are *nascently moral*: They experience moral emotional responses and empathic distress when witnessing animals being harmed; they quickly and spontaneously report harms to parents and teachers; and they can recognize animal needs and act to fulfill those needs, perhaps even adjusting their own needs to compromise with the needs of the animal other. As early as preschool, children perceive animals as beings who display the hallmarks of experiencing subjects (Melson, 2001; Myers, 2007; Myers & Saunders, 2002). The child softly pets his or her puppy and the puppy nestles deeper in the child's arms, but when the child pulls the animal's tail the animal runs away. The child runs toward the flock of pigeons and the birds take flight. The child probes the insect and it curls into a tightly wound ball. The child learns of the animal as a subjective experiencing other with needs and experiences not so similar to, but perhaps also not so different from, their own.

Moral and folkbiological literature merges on a few key points that give ground to the argument that the findings of the current study can improve early education that affects not only young and middle childhood but also preschool-age children. The biological constructs present in the moral thinking of the young children in the current study are observed in kids as young as preschool age (see Inagaki & Hatano, 2002, for a review). Preschoolers understand the living status of animals (e.g., Huang & Lee, 1945) and the biological needs and processes (e.g., Inagaki & Hatano, 1996). Preschoolers are concerned with the physical welfare of others, including animals (Zelazo et al., 1996), and will respond in prosocial and moral ways to harms done to animals (Myers, 2007). The building blocks of the moral regard observed in the young children in the current study are seen emerging in research on preschool-age children. Educators can seek to provide preschoolers' with opportunities to encounter and engage these building blocks so that children have them in place as they move forward in moral development.

Implications for policy and education

Children are the next generation of conservationists, and they will have a huge task on their hands when it comes to the preservation of biodiversity. It is estimated that 30,000 species go extinct each year (Wilson, 2006). To support children's capacity to engage in conservation psychologists and educators need to support their developing moral concerns for the environment. The findings from the current study highlight that children think about the welfare and rights of endangered animals in deep and beautiful ways. These findings could shape the conservation agenda by informing environmental education and policy that prepares children to be conservationists.

By synthesizing an account of children's patterns of thought that support conservation concerns (e.g., children's tendency to draw on biological needs and capabilities to ground their claims for protecting a species), the current study provides a toolkit for educators to use. The findings could support attuning teachers to recognizing what is happening in children's ideas, recognizing what frameworks children are working with, and building from these to support learning. Educators could hone their abilities to recognize patterns in children's thinking and to employ salient features of the environmental context that can spark the construction of knowledge about the environment, particularly the forms of knowledge that support moral thinking and conservation concerns.

More specifically, the findings from the current study could be applied in science education to improve a curriculum that will support children's understanding of ecological relationships. One of the current tasks in environmental education (Medin & Bang, 2014) is to teach teachers how to use the language of plant and animal capabilities—for example, how to teach children about the ways in which plants communicate in response to cues from natural stimuli in the environment. These lessons support the ways that children comprehend relationship dynamics between humans, animals and plants, and the wider natural world. Understanding the capabilities of one entity supports a more complicated understanding of wider ecological relations. For example, wolves' hunting behavior will affect ungulates' (e.g., deer) feeding behavior, which will impact the plant growth in a given area, and such processes have been shown to then further impact flora growth and ecological topography, such as the movement of rivers (see Eisenberg, 2010). The findings of the current study show the quality and the import of children's ecological construals in shaping moral concern for the wider natural world. The patterns observed in the findings can act as a blueprint for developing class material and activities that support the work of teaching children about ecological relations.

The current study could be employed to inform environmental policy decisions. Specifically, policymakers might rethink the ways in which they frame their decision making, drawing more deeply on human values, including children's values, of the natural world to inform policy decisions. The current study offers insight into the perspective of children who care about and value endangered animals. The findings highlight that the issue of endangered species matters to them. By drawing on children's values in the development of environmental policy psychologists and educators engage them in the policy process. It is imperative that psychologists and educators empower children to recognize their potential to affect the world in meaningful ways. Consider the words of one student who was part of a group of students rallying for control to determine the use of a 2.3-acre plot of land in Keene, New Hampshire: "Your decision to involve us or not will determine our future commitment to this town and to public decision making. 'It will effect whether we think we can really make a difference" (Thomashow, 2002, p. 270). If psychologists and educators do not engage children in the matters of environmental policy and conservation, they may not engage themselves.

References

Almeida, A., Vasconcelos, C. M., Strecht-Ribeiro, O., & Torres, J. (2013). Non-anthropocentric reasoning in children: Its incidence when they are confronted with ecological dilemmas. *International Journal of Science Education*, *35*(2), 312–334. doi:10.1080/09500693.2011.608387

Atran, S. (1998). Folk biology and the anthropology of science: Cognitive universals and cultural particulars. *Behavioral & Brain Sciences*, *21*, 547–609. doi:10.1017/S0140525X98001277

Backscheider, A. G., Shatz, M., & Gelman, S. A. (1993). Preschoolers' ability to distinguish living kinds as a function of regrowth. *Child Development, 64*, 1242–1257. doi:10.2307/1131337

Berenguer, J. (2010). The effect of empathy in environmental moral reasoning. *Environment and Behavior, 42*, 110–134. doi:10.1177/0013916508325892

Carey, S. (1985). *Conceptual change in childhood.* Cambridge, MA: MIT Press.

Clayton, S., & Myers, O. E., Jr. (2009). *Conservation psychology: Understanding and promoting human care for nature.* New York, NY: Wiley/Blackwell.

Coley, J. D. (1995). Emerging differentiation of folkbiology and folkpsychology: Attributions of biological and psychological properties to living things. *Child Development, 66*, 1856–1874. doi:10.2307/1131915

Coley, J. D. (2000). On the importance of comparative research: The case of folkbiology. *Child Development, 71*, 82–90. doi:10.1111/1467-8624.00121

Damon, W. (1977). *The social world of the child* (pp. 137–139). San Francisco, CA: Jossey-Bass.

Eisenberg, C. (2010). *The wolfs tooth: keystone predators, trophic cascades, and biodiversity.* Washington, DC: Island Press.

Fleiss, J. L., Levin, B., & Paik, M. C. (2003). *The standardization of rates. Statistical Methods for Rates and Proportions.* In (3rd ed., pp. 627–647).

Gelman, R., & Spelke, E. S. (1981). The development of thoughts about animate and inanimate objects: Implications for research on social cognition. In J. H. Flavell (ed.), *Social cognitive development: Frontiers and possible futures* (pp. 43–66). Cambridge, UK: Cambridge University Press.

Grace, M. M., & Ratcliffe, M. (2002). The science and values that young people draw upon to make decisions about biological conservation issues. *International Journal of Science Education, 24*(11), 1157–1169. doi:10.1080/09500690210134848

Greif, M. L., Nelson, D. G. K., Keil, F. C., & Gutierrez, F. (2006). What do children want to know about animals and artifacts? Domain-specific requests for information. *Psychological Science, 17*, 455–459. doi:10.1111/j.1467-9280.2006.01727.x

Hatano, G., & Inagaki, K. (1994). Young children's naïve theory of biology. *Cognition, 50*, 1–25. doi:10.1080/03004430.2011.636811

Helwig, C. C. (1995). Adolescents' and young adults' conceptions of civil liberties: Freedom of speech and religion. *Child Development, 66*, 152–166. doi: 10.1111/j.1467-

Hickling, A. K., & Gelman, S. A. (1995). How does your garden grow? Early conceptualization of seeds and their place in the plant growth cycle. *Child Development, 66*, 856–876. doi:10.2307/1131955

Huang, I., & Lee, H. W. (1945). Experimental analysis of child animism. *The Pedagogical Seminary and Journal of Genetic Psychology, 66*, 69–74. doi:10.1080/08856559.1945.10533311

Inagaki, K. (1990). The effects of raising animals on children's biological knowledge. *British Journal of Developmental Psychology, 8*, 119–129. doi:10.1111/j.2044-835X.1990.tb00827.x

Inagaki, K., & Hatano, G. (1993). Young children's understanding of the mind-body distinction. *Child Development, 64*, 1534–1549. doi:10.2307/1131551

Inagaki, K., & Hatano, G. (1996). Young children's recognition of commonalities between animals and plants. *Child Development, 67*, 2823–2840. doi:10.2307/1131754

Inagaki, K., & Hatano, G. (2002). *Young children's naïve thinking about the biological world.* New York, NY: Psychology Press.

Kahn, P. H., Jr. (1992). Children's obligatory and discretionary moral judgments. *Child Development, 63*, 416–430. doi:10.2307/1131489

Kahn, P. H., Jr. (1999). *The human relationship with nature: Development and culture.* Cambridge, MA: MIT Press.

Kahn, P. H., Jr., & Lourenço, O. (2002). Water, air, fire, and earth: A developmental study in Portugal of environmental moral reasoning. *Environment and Behavior, 34*, 405–430. doi:10.1177/00116502034004001

Kahn, P. H., Jr., Saunders, C. D., Severson, R. L., Myers, O. E., Jr., & Gill, B. T. (2008). Moral and fearful affiliations with the animal world: Children's conceptions of bats. *Anthrozoös, 21*, 375–386. doi:10.2752/175303708X371591

Kahn, P. H., Jr., Severson, R. L., & Ruckert, J. H. (2009). The human relation with nature and technological nature. *Current Directions in Psychological Science, 18*, 37–42. doi:10.1111/j.1467-8721.2009.01602.x

Karpiak, C. P., & Baril, G. L. (2008). Moral reasoning and concern for the environment. *Journal of Environmental Psychology, 28*, 203–208. doi:10.1016/j.jenvp.2007.12.001

Katcher, A. (2002). Animals in therapeutic education: Guides into the liminal state. In P. H. Kahn, Jr., & S. R. Kellert (Eds.), *Children and nature: Psychological, sociocultural, and evolutionary investigations* (pp. 180–198). Cambridge, MA: MIT Press.

Keil, F. C. (1986). The acquisition of natural kind and artifact terms. In W. Demopoulos and A. Marras (Eds.), *Language Learning and Concept Acquisition* (pp. 133–153).

Keil, F.C. (1995). The growth of causal understandings of natural kinds: Modes of construal and the emergence of biological thought. In. A. Premack and D. Sperber (Eds.), *Causal Cognition.* Oxford: Oxford University Press.

Kelemen, D. (1999). Why are rocks pointy? Children's preference for teleological explanations of the natural world. *Developmental Psychology, 35*, 1440–1452. doi:10.1037/0012-1649.35.6.1440

Kellert, S. R. (1996). *The value of life*. Washington, DC: Island Press.

Kellert, S. R. (2002). Experiencing nature: Affective, cognitive, and evaluative development in children. In P. H. Kahn, Jr., & S. R. Kellert (Eds.), *Children and nature: Psychological, sociocultural, and evolutionary investigations* (pp. 117–152). Cambridge, MA: MIT Press.

Kohlberg, L. (1984). *Essays on moral development: The psychology of moral development* (vol. 2). San Francisco, CA: Harper & Row.

Medin, D. L., & Atran, S. (2004). The native mind: Biological categorization and reasoning in development and across cultures. *Psychological Review, 111*, 960–983. doi:10.1037/0033-295X.111.4.960

Medin, D. L., & Bang, M. (2014). *Who's asking? Native science, Western science, and science education*. Cambridge, MA: MIT Press.

Melson, G. F. (2001). *Why the wild things are: Animals in the lives of children*. Cambridge, MA: Harvard University Press.

Myers, O. E., Jr. (2007). *The significance of children and animals: Social development and our connections to other species* (2nd rev. ed.). West Lafayette, IN: Purdue University Press.

Myers, O. E., Jr., & Saunders, C. (2002). Animals as links to developing caring relationships with the natural world. In P. H. Kahn, Jr., & S. R. Kellert (Eds.), *Children and nature: Psychological, sociocultural and evolutionary investigations* (pp. 153–178). Cambridge, MA: MIT Press.

Nevers, P., Gebhard, U., & Billmann-Mahecha, E. (1997). Patterns of reasoning exhibited by children and adolescents in response to moral dilemmas involving plants, animals and ecosystems. *Journal of Moral Education, 26*, 169–186. doi:10.1080/0305724970260205

Nucci, L. P. (2001). *Education in the moral domain*. Cambridge, UK: Cambridge University Press.

Office of Financial Management. (2013, March). *State of Washington Median Household Income Estimates by County: 1989 to 2013 and Projection for 2014*. Retrieved March 6, 2013 from http://www.ofm.wa.gov/economy/hhinc/medinc.pdf

Paltved-Kaznelson, C. (2009). *Nature through the generations*. The Danish Society for Nature Conservation. The National Survey of Denmark Gallup. Retrieved from http://www.dn.dk/Default.aspx?ID=4592

Piaget, J. (1932). *The child's conception of the world*. Savage, MD: Littlefield Adams Quality Paperbacks.

Poling, D. A., & Evans, E. M. (2004). Are dinosaurs the rule or the exception? *Cognitive Psychology, 19*, 363–383. doi:10.1016/j.cogdev.2004.04.001

Resnick, J., Stickgold, R., Rittenhouse, C. D., & Hobson, J. A. (1994). Self-representation and bizarreness in children's dream reports collected in the home setting. *Consciousness and Cognition, 3*, 30–45. doi:10.1006/ccog.1994.1003

Rideout, V. J., Foehr, U. G., & Roberts, D. F. (2010). *Generation M2: Media in the lives of 8-to 18-year-olds*. Henry J. Kaiser Family Foundation. Retrieved March 6, 2013 from http://files.eric.ed.gov/fulltext/ED527859.pdf

Ross, N., Medin, D. L., Coley, J. D., & Atran, S. (2003). Cultural and experiential differences in the development of biological induction. *Cognitive Development, 18*, 25–47. doi:10.1016/S0885-2014(02)00142-9

Severson, R. L., & Kahn, P. H., Jr. (2010). In the orchard: Farm worker children's moral and environmental reasoning. *Journal of Applied Developmental Psychology, 31*, 249–256. doi:10.1016/j.appdev.2010.02.003

Smetana, J. G. (1981). Preschool children's conceptions of moral and social rules. *Child Development, 52*, 1333–1336. doi:10.2307/1129527

Springer, K., & Keil, F. C. (1991). Early differentiation of causal mechanisms appropriate to biological and non-biological kinds. *Child Development, 62*, 767–781. doi:10.2307/1131176

Taylor, M., & Carlson, S. M. (1997). The relation between individual differences in fantasy and theory of mind. *Child Development, 68*, 436–455. doi:10.2307/1131670

Thomashow, C. (2002). Adolescents and ecological identity: Attending to wild nature. In P. H. Kahn, Jr., & S. Kellert (Eds.), *Children and nature: Psychological, sociocultural, and evolutionary investigations* (pp. 259–278). Cambridge, MA: MIT Press.

Turiel, E. (1983). *The development of social knowledge: Morality and convention*. Cambridge, UK: Cambridge University Press.

Turiel, E., & Davidson, P. (1986). Heterogeneity, inconsistency, and asynchrony in the development of cognitive structures. In I. Levin (Ed.), *Stage and structure: Reopening the debate* (pp. 106–143). Norwood, NJ: Ablex Press.

Vosniadou, S., & Brewer, W. F. (1992). Mental models of the earth: A study of conceptual change in childhood. *Cognitive psychology, 24*, 535–585. doi:10.1016/0010-0285(92)90018-W

Wainryb, C. (1995). Reasoning about social conflicts in different cultures: Druze and Jewish children in Israel. *Child Development, 66*, 390–401. doi:10.2307/1131585

Waxman, S., & Medin, D. (2007). Experience and cultural models matter: Placing firm limits on childhood anthropocentrism. *Human Development, 50*, 23–30. doi:10.1159/000097681

Wilson, E. O. (2006). *The creation: An appeal to save life on earth*. New York, NY: Norton.

Zelazo, P. D., Helwig, C. C., & Lau, A. (1996). Intention, act, and outcome in behavioral prediction and moral judgment. *Child Development, 67*, 2478–2492. doi:10.2307/1131635

Direct Experience With Nature and the Development of Biological Knowledge

Sarah E. Longbottom and Virginia Slaughter

ABSTRACT
Research Findings: An emerging consensus is that casual, direct contact with nature influences the development of children's biological knowledge. Here we review the existing literature on this topic, focusing on the effects of (a) rural versus urban rearing environments and (b) pet ownership and care on children's biological concepts and reasoning. Although the research is limited, the evidence suggests that these factors positively influence children's understanding of specific biological phenomena (e.g., internal organ function, biological inheritance) and reasoning patterns (e.g., anthropocentrism, ecological reasoning). *Practice or Policy*: In the modern context of diminishing outdoor playtime and limited experience with nature, this review highlights the importance of direct exposure to the natural world for children's conceptual development in the domain of biology.

Considering that many children today spend more time indoors than outdoors (Planet Ark, 2013), exploring how informal, personal experience with nature influences cognitive development is vital from basic research and applied perspectives. A small body of research conducted over the past 20 years suggests that direct, everyday experience with nature positively influences young children's factual knowledge and reasoning in the domain of biology. Here we review the relevant findings in order to draw some broad conclusions and to highlight directions for further research.

We begin by outlining some of the reasons why this topic is important in the modern context. Then we review studies that have examined the development of biological concepts with respect to (a) rural versus urban upbringings and (b) pet ownership and care. We end with a summary discussion that includes areas of future exploration and potential applications of the findings to date.

Children and Nature

The amount of time children spend outdoors and interacting with nature is declining in many Western societies (Malone, 2007). For instance, Planet Ark (2013) conducted an online survey of 1,002 Australians ages 14–64 years with demographic characteristics generally representative of the national Australian population. The survey was designed to investigate children's interaction with nature and participation in outdoor activities, with parents of children younger than 16 years reporting on their child's activities. The study found that over the course of an entire month in late summer, only 37% of surveyed children had played outdoor games in a backyard or playground and only 39% had visited a park or reserve. Furthermore, only 35% of surveyed children played outside every day, and just 13% spent more time playing outside than indoors. Another survey by Planet Ark (2011) exploring children's outdoor activities in detail revealed that in the 12 months prior to the survey, just 54% of children had gone bushwalking, 33% had gone camping, and fewer than 60% had visited a national park. The survey further revealed that a substantial minority of children up to age 16 had missed out entirely on some outdoor experiences: Specifically, 32% of children had never been camping, 29% had never been bushwalking, 27% had never climbed a tree, 28% had never cared for a vegetable garden, and 31% had never planted or cared for a tree or shrub. The survey also indicated that on average Australian children spent 5.1 hr per week playing outdoors, with this overall average slightly higher for rural children and lower for urban children (Planet Ark, 2011).

A decrease in time spent outdoors in early childhood over the past 20 years has also been observed internationally (see Charles & Wheeler, 2012, for more information). Studies in the United States have found small but significant declines in children's participation in outdoor activities over the past two decades (Charles & Wheeler, 2012). Highlighting this trend, a recent study in the United Kingdom comparing urban children's ability to identify native British wildlife and Pokémon characters found that children 6 years and older could identify more Pokémon than real, locally found animals and plants (Balmford, Clegg, Coulson, & Taylor, 2002). This finding is consistent with studies suggesting that diminished contact with, and understanding of, nature is associated with more developed, technologically oriented societies (Atran, Medin, & Ross, 2004).

Understanding how contact with plants and animals influences children's knowledge acquisition is especially vital in light of recent studies examining *where* young children have reported learning biological knowledge. For example, Patrick, Tunnicliffe, and their colleagues (Patrick et al., 2013; Patrick & Tunnicliffe, 2011; Tunnicliffe & Reiss, 1999) have found relatively consistent trends in America, Brazil, England, Finland, Iceland and Portugal in 4- to 15-year-olds' sources of knowledge about animals. In these studies children cited their main sources of information about animals as informal firsthand experiences at home, with friends, in the garden, and on the way to school. The media (books and television) and the zoo were cited as intermediate sources, whereas school reportedly contributed little to their knowledge base. A similar trend was observed in parallel studies examining plant knowledge (Patrick & Tunnicliffe, 2011; Tunnicliffe & Reiss, 2000). Based on these findings, Patrick and Tunnicliffe (2011) concluded that "school does not have a considerable influence on how children understand objects in the natural world, especially at younger ages" (p. 640), reflecting Shepardson's (1997) suggestion that casual, daily experiences are more directly associated with children's initial understandings of science than formal learning.

Contact with nature and animals in childhood also influences people's attitudes toward nature and conservation, their interest in these topics, and their future career choices (Charles & Wheeler, 2012; Prokop, Prokop, & Tunnicliffe, 2008; Prokop, Prokop, Tunnicliffe, & Diran, 2007; Zhang, Goodale, & Cheng, 2014). For example, a recent study in Finland (Uitto, Juuti, Lavonen, & Meisalo, 2006) found that teenagers who engaged in nature-based activities outside of school, such as bushwalking, also had strong interests in applied biology and zoology, and participation in farm-based activities was associated with greater interest in applied biology. Similarly, a study by the Australian Council for Educational Research (2002) found that teenagers who lived in rural areas of Australia were more inclined to study health, home science, and agriculture in their final year of high school than their urban counterparts.

These findings point to direct experience with nature as an important factor in children's developing interest in, and knowledge about, the biological world. Although the research literature to date lacks a consensus definition, direct experience with nature is typically conceptualized as an individual's informal, firsthand interactions with the plants and animals in his or her everyday life. This is distinguished from formal teaching about biology, although these may overlap at times, for instance on school field trips. In the literature we review here on children's developing understanding of biology, direct experience has been operationalized as (a) rearing environment, with a focus on contrasting urban versus rural upbringings; and (b) pet ownership and care. Biological knowledge in this literature refers to children's intuitive, folkbiological concepts and reasoning about the natural world (Carey, 1985). This has been assessed with a variety of cognitive tasks encompassing the acquisition of specific biological facts, predictions about plant and/or animal physiology and ecology, and inductive reasoning about living things.

Rearing Environment

On the assumption that children who grow up in rural areas have greater opportunities to observe and interact with nature than their urban counterparts, numerous studies have compared urban versus rural children's biological knowledge. The idea is that everyday contact with plants and animals in "relatively intact ecosystems" (Coley, 2012, p. 994) provides rich information from which children acquire facts, notice patterns, and make inferences about the natural world. Although it appears self-evident, we are aware of only two studies that have actually tested the assumption that rural children have more experience with nature than urban children. Zhang et al. (2014) and Coley (2012) both examined the role of rural versus urban rearing environment on children's experiences with nature. These two studies were conducted in different populations, namely, China and North America, respectively, but produced similar findings indicating that the frequency of nature-based activities decreased as urbanization and population density increased. These findings support the assumption that children from rural areas have more direct experience with nature.

Anthropocentric Reasoning

A central element of learning and reasoning about biology is understanding the scope of the domain, namely, all living things. In one of the foundational studies of the development of biological reasoning, Carey (1985) developed a biological property projection task in which children were taught about a novel property of a living thing (i.e., dogs have axons) and asked whether other living things (e.g., birds, fish, humans) share that property. Carey's key finding, obtained from testing urban American children, was that anthropocentric reasoning, whereby humans have privileged status, was the norm in early childhood. Specifically, young children reason anthropocentrically when they assert that properties found in humans are shared by other animals but not vice versa. Carey argued that around ages 8–10, children's thinking about the biological world undergoes a fundamental conceptual change, which gives rise to a mature framework in which humans are seen as just one type of animal with largely the same biological properties as other living things.

Carey's (1985) conclusion has been repeatedly challenged by researchers concerned that urban and rural populations may differ in the development of biological reasoning. For instance, Ross, Medin, Coley, and Atran (2003) explored the effects of experience and culture on young children's performance in a version of Carey's task. They compared the reasoning patterns of urban and rural Euro-Americans and rural Native Americans from the Menominee tribe, all ages 6–10 years. They found that the urban children showed strong anthropocentric reasoning at all ages, whereas the Menominee children showed no sign of anthropocentric reasoning at any age. The rural Euro-Americans showed some signs of anthropocentrism at the younger ages, but the trend disappeared with age. This finding suggests that rural children's reasoning about the domain of biology is more sophisticated than that of their urban peers, and furthermore culture (e.g., Menominee vs. Euro-American) may be an added influence.

Similarly, Waxman and Medin (2007) compared urban and rural 4- to 7-year-olds' performance on a slightly different property projection task and found an anthropocentric reasoning pattern only in the youngest urban children. The authors favored the conclusion that anthropocentrism is only characteristic in the biological reasoning of young urban children, but they also conceded that methodological differences relative to Carey's (1985) procedure could have accounted for the weak indications of anthropocentric reasoning.

Medin, Waxman, Woodring, and Washinawatok (2010) then replicated Carey's (1985) original methodology. They sought to clarify whether rearing environment genuinely influences biological reasoning patterns in young children or whether the discrepancies between previous studies and Carey's findings reflect methodological differences. They compared the reasoning patterns of children ages 4–10 years recruited from the same populations used in Ross et al.'s (2003) study (urban and rural Euro-American and rural Menominee children) and found that the urban children showed similar patterns of anthropocentric reasoning to those reported by Carey. However, again, a distinctly different pattern emerged in the rural populations. Even the youngest rural Euro-Americans showed no signs of anthropocentrism. The differences between urban and rural children's reasoning were most pronounced in the youngest age group, a finding that has also been reported in studies comparing urban and traditional cultures (see, e.g., Waxman, Medin, & Ross, 2007). Overall these findings demonstrate that a rural rearing environment is associated with more advanced, non-anthropocentric reasoning about the biological world in early childhood.

Building on this work, Tarlowski (2006) explored the dual factors of rearing environment and parental transmission of knowledge on young children's patterns of biological property projection. This study included an expanded version of Carey's (1985) task in which nonliving things (e.g., dolls, rocks, pillows) were also included as targets. Forty-five 4- and 5-year-old Polish children were recruited based on two factors: (a) where they lived (urban or rural) and (b) their parent's biological expertise (expert or layperson). This study revealed that both rearing environment and parental expertise significantly influenced children's biological reasoning. Specifically, rural children made fewer property projections overall than their urban counterparts, thereby displaying greater sensitivity to the division between living and nonliving targets. At the same time, experts' children tended to make projections based on biological categories, for instance attributing a human property to other mammals, whereas the children of laypeople relied more on visual similarity (e.g., attributing a human property to a doll).

These findings further support the proposal that rural rearing environments are associated with more sophisticated biological reasoning in children, and they add the suggestion that parental transmission of information about the biological world may also be a significant influence. However, Tarlowski (2006) acknowledged that without detailed analyses of the children's everyday experiences with nature and features of parental input about biology, it is premature to conclude that those specific factors were responsible for the pattern of findings. For instance, it may be that rather than directly transmitting information about the natural world to their children, biology expert parents structure everyday experiences that increase their children's direct and indirect exposure to nature, for instance by favoring family bushwalking over watching movies or choosing nature television over cartoons. Along the same lines, rural children's direct exposure to nature may positively influence their precocious reasoning about biology, but it may also be the case that rural children experience parental, peer, and media input that differs from that of their urban counterparts and that these indirect sources of information are driving the effect.

Ecological Reasoning

Ross et al. (2003) encountered an unexpected reasoning strategy in rural Euro-American and Menominee children. They found that these children tended to refer to ecological relations when justifying their responses on the property projection task. For instance, they might have claimed that bees and bears share a property because bears eat honey.

Although their task had not been designed to measure ecological reasoning, Ross et al. (2003) took the opportunity to assess whether rearing environment was associated with different patterns of reasoning with respect to ecological relations. They found a reliable interaction between age and rearing environment, with ecological reasoning evident in the older rural Euro-American children and in all of the Menominee children regardless of age. This ecological mindset was not observed in urban children of any age. These findings extend the conclusion that a rural upbringing influences children's biological reasoning, being associated with justifications that refer to natural, ecological relations between living things as well as relatively mature patterns of property projection. Given that the Menominee and Euro-American children from similar rural environments showed different developmental patterns of ecological reasoning and property projection, Ross et al. concluded that "both culture and experience play an important role" (p. 45) in shaping young children's thinking about biology. Thus, some aspect of Menominee life seems to have led those children to eschew anthropocentrism at all ages while embracing ecological reasoning. Ross et al. speculated that Menominee practices of gathering traditional wild foods and fishing sustainably may draw children's attention to similarities and interrelations between different species in their local environment.

Coley (2012) and Coley, Vitkin, Seaton, and Yopchick (2005) further investigated the frequency of ecological reasoning in children from different rearing environments. They developed a triad induction task to pit ecological and taxonomic reasoning against each other. Children were taught a novel property about a base living thing and then asked which of two targets—one that was ecologically related (e.g., shared habitat) and the other that was taxonomically related (e.g., a related species)—would share that property with the base. This design enabled them to build on previous findings from adult laypeople and biology experts that showed that when projecting an internal property, both groups tended to use taxonomic relations, but when projecting a disease, experts tended to use ecological relations whereas novices continued to favor taxonomic reasoning (Shafto & Coley, 2003). Thus, Coley et al. and Coley speculated that because of their greater experience with nature, rural children would show similar reasoning patterns to those of the adult experts. As predicted, both studies revealed significant effects of rearing environment on children's reasoning: Rural children made significantly more ecological projections than urban children, with suburban children falling between the two groups and not differing significantly from either. Furthermore, children's patterns of reasoning about different properties reflected those seen in adult experts and novices, such that suburban and rural children made significantly more ecological projections when reasoning about diseases, but no significant differences existed between any of the groups in reasoning about internal properties. As a further test of the effect of rearing environment, the authors conducted multiple regressions using population density as a predictor. These analyses confirmed that the frequency of ecological projections increased as population density decreased (i.e., became more rural). Finally, Coley et al. found that children made significantly more ecological projections when reasoning about local species, with which they presumably had more direct experience, than exotic species. This finding adds weight to the idea that rearing environment provides children with direct experience that in turn promotes more sophisticated understandings of biology.

Knowledge About Biological Inheritance and Internal Organs

Complementing this research on patterns of reasoning about living things, a couple of studies have evaluated the influence of rearing environment on children's knowledge about specific biological concepts. For instance, Williams and Smith (2006) found an effect of rearing environment on children's understanding of biological inheritance. Using an adoption task in which children were asked to reason about the traits and behaviors an adopted baby would share with her parents, Williams and Smith found that rural children displayed a more advanced understanding of parents' contributions to biological traits than urban children. Furthermore, they found that rural children offered more nuanced explanations of social and biological aspects of inheritance than their urban and suburban peers.

Panagiotaki and Nobes (2014) compared British Muslim, British non-Muslim, and Pakistani Muslim children's knowledge about the role of vital organs in the maintenance of life. They reasoned that any differences between the two British groups and the Pakistani Muslim children would reflect the effects of contact with animals, as the British samples grew up in similar urban environments whereas the Pakistani children were raised in a rural village where animals were an important part of everyday life. They found that despite a lack of awareness of the specific functions of internal organs, the Pakistani children showed a significantly greater understanding of the importance of organs for staying alive compared to their British counterparts.

The Role of Rearing Environment: Summary

All together, the research comparing rural and urban children's reasoning patterns and knowledge about specific biological concepts indicates that a rural upbringing is associated with relatively advanced understanding. There is some suggestion in the data to support the proposal that a rural upbringing entails everyday experiences that engender expertise about the local biological world, and this drives the rural versus urban differences observed across the various studies. However, it is not yet clear what those formative everyday experiences are: One assumption is that direct contact with nature is the crucial variable; however, it is also possible that more indirect influences such as input from parents, peers, different school curricula, and/or media could drive the observed differences or at least play a role. Rural life differs from urban and suburban life on numerous dimensions other than informal, direct experience of the natural environment, and these confounded factors have rarely been considered separately. One such factor that has itself been a focus of research is pet ownership.

Keeping Pets

The beneficial effects of keeping pets on children's and adults' well-being, social skills, and physical and mental health have been well established (Friedman & Son, 2009; Muldoon, Williams, Lawrence, Lakestani, & Currie, 2009). These benefits of pet ownership have been popular research topics; however, researchers have only recently begun to explore whether pet ownership in early childhood influences the acquisition of biological knowledge and concepts. Whereas the majority of studies assessing the role of rearing environment have focused on children's reasoning about biological relations (using property induction tasks as described previously), here the focus is primarily on children's understanding of specific biological concepts and phenomena.

Pet Ownership

Prokop et al. (2007) explored the effects of owning pets on children's knowledge of the internal structures (e.g., organs and skeletons) of vertebrates and invertebrates. They used a free drawing task in which children were asked to portray the insides of different animals that were presented in photos or as stuffed specimens. Their first study, involving 702 Slovak children ages 6–16 years, revealed more detailed knowledge about internal organs among the subset of children who had a pet; however, the effect was just nonsignificant ($p = .06$). In contrast to some other research (e.g., Patrick et al., 2013; Patrick & Tunicliffe, 2011), the children in this study cited school books as sources of information about animal biology more frequently than they cited personal experience. In a follow-up study with 1,544 Slovak 6- to 15-year-olds, Prokop et al. (2008) again found that children who raised pets had significantly more detailed knowledge of the internal structure of a variety of animal species compared to children who did not own a pet. Furthermore, this knowledge was significantly greater among children who owned more than two pets compared to children who raised a single pet or did not own a pet at all.

Recent research by Prokop and Tunnicliffe (2010) extended this finding by showing that the number of pets a child owns is also linked to his or her knowledge of animal ecology. These researchers examined the influence of pet ownership on children's attitudes toward, and knowledge about, a variety of animals that varied by type (e.g., pest, disease carrier, and predator) and popularity in Slovak society. In three separate studies, pet-owning 10- to 15-year-olds generally displayed more detailed knowledge of, and held more positive attitudes toward, each animal compared to their peers who did not own a pet. Replicating Prokop et al. (2008), the researchers found an effect of number of pets, whereby children who owned three or more pets displayed the greatest degree of knowledge about each type of animal (with the exception of the squirrel) and the most positive attitudes toward both the popular and unpopular animals.

In a more recent study, Geerdts, Van de Walle, and LoBue (2015a) assessed associations between pet ownership and children's biological knowledge using a property projection task (Carey, 1985) and a free-response interview asking about the biological properties and psychological attributes of a pet. This study revealed that 3- to 5-year-old children with pets were less likely to exhibit anthropocentric reasoning in the property projection task than their non-pet-owning counterparts, which is consistent with the research contrasting rural and urban children discussed previously. Children with pets were also more likely to attribute biological properties to pets compared to children without pets. Based on direct observation of children interacting with their pets and parent reports, these researchers drew the conclusion that the majority of these children's interactions with their pets were social exchanges rather than care responsibilities. However, this goes against other research suggesting that caring for pets in particular has a positive influence on children's developing biological knowledge.

Caring for Pets

The positive influence of pet care was first documented by Inagaki's (1990) pioneering study examining the effect of raising goldfish on Japanese children's biological knowledge. Kindergartners who owned and cared for goldfish were compared with a control group who had never owned or raised any animal. Both groups of children answered factual knowledge questions about goldfish and mammals and made predictions about biological phenomena relating to goldfish, people, and frogs. Analyses revealed that the goldfish-raising group possessed more factual knowledge about goldfish, made more reasonable predictions about goldfish, and made significantly more goldfish analogies when making predictions about frogs compared to the control group. This pattern of findings was replicated in a study focusing on mammal pets instead of goldfish (Inagaki, cited in Hatano & Inagaki, 2013).

The role of pet care was assessed by Toyama, Lee, and Muto (1997) in the context of classroom pets. In the Tokyo school where they conducted the study, children were classified as those who regularly engaged in classroom pet care procedures versus those children who rarely or never chose to care for the animals. These two groups were asked a series of questions relating to their caregiving activities and made predictions about what would happen if pet care procedures went unfulfilled. Results indicated that children who regularly cared for the animals made more than 3 times as many biological statements while performing care procedures (12.8% compared to 3.6% for noncarers) and were significantly more likely to provide biological justifications for caregiving procedures, for example explaining that they cut up food for the ducks because ducks do not have teeth. Finally, the pet carers were also more accurate in predicting the biological implications of failing to carry out appropriate care procedures.

Myers, Saunders, and Garrett (2004) assessed relations between pet ownership and children's understanding of animals' needs, including environmental and conservation requirements for individual and species survival. They asked 171 children ages 4–14 years who were visiting a local zoo to answer questions about the various needs of an animal nominated by the child. Frequency of zoo visits, pet ownership, pet care responsibilities, and an interest in animals were also assessed. This research revealed that the more pets a child owned, the greater his or her understanding of animals' environmental and conservation (e.g., protective legislation) needs. Highlighting the role of pet care, the researchers found that children who reported having pet care responsibilities obtained higher environmental, conservation, and total needs scores compared to children who did not care for pets.

Significant Biological Events in the Pet Life Cycle

Although there is some empirical support for the idea that hands-on pet care is important, the only intervention study to date suggests that simply witnessing natural biological life events in pet animals can promote children's biological reasoning. For this study, Herrmann, French, DeHart, and

Rosengren (2013) recruited an experimental group of 30 children ages 3 and 4 years old in a North American preschool and a smaller (n = 14) control group from a different preschool. They presented all children with an initial baseline test assessing their willingness to attribute biologically relevant versus impossible changes to a variety of animals. They next implemented an intervention for the experimental group in which two terrariums, containing a pregnant mouse and a pair of caterpillars, respectively, were placed in a room adjoining the children's classrooms. Over 2 weeks the children observed both terrariums, drew pictures of events within the terrariums, and asked their teachers questions about those events, although no formal instruction took place. By the end of the 2-week intervention, the mouse had given birth to pups and the caterpillars had gone through all of the stages of metamorphosis, after which the children released the butterflies. All children were given a posttest that revealed that the experimental group was more accurate in attributing life span changes to animals, particularly when asked about caterpillars. No significant alteration in performance from pre- to posttest was observed in the control group. These results indicate that observation of pet animals—and not necessarily direct caregiving experience—leads to an increase in biological knowledge, although the special significance of witnessing metamorphosis and/or birth is unknown, as there was no comparison group who had access to the animals when they were not undergoing biological transformational events. However, the result of this intervention study is in line with Williams and Smith's (2006) report that those children whose pets had given birth possessed a more sophisticated understanding of biological inheritance than their peers whose pets had not given birth. It is noteworthy that in both of these studies, children's knowledge about specific biological phenomena matched up to their direct experience, in line with Herrmann et al.'s argument for "the necessity of real-world experience with living kinds in order for certain beliefs about biological kinds to manifest" (p. 216).

The Role of Pet Ownership and Care: Summary

The research suggests that regular exposure to pet animals in childhood, either at home or in school, is associated with relatively sophisticated biological concepts. Based on the literature to date it is not yet clear whether this association is driven by the experience of caring for the animal, by social interaction with the animal, or simply by the animal's presence. These data also suggest that there is a positive correlation between the number of pets in children's lives and their level of biological understanding. This effect may be a function of varied experiences with the pet; indeed, there are also data to suggest that witnessing a significant biological event (birth, metamorphosis) is particularly influential in promoting children's biological knowledge. However, there are too few studies isolating different types of direct experience with pets to draw firm conclusions, so carefully designed future studies will be required to understand the underlying mechanisms.

Conclusions, Open Questions, and Future Directions

Table 1 lists the methodological details of all of the studies reviewed here and their findings. Taken together, the results of these studies support the broad conclusion that informal, direct exposure to nature has a positive influence on children's developing understanding of biology. The research to date, although limited, has some strengths. One of these is geographical diversity: The positive findings are reasonably consistent across the United States, the United Kingdom, Eastern Europe, and Japan. It may also be considered a strength that biological knowledge has been assessed with a variety of tasks, although, as discussed below, this makes it hard to draw general conclusions.

Across the studies reviewed, exposure to nature was assessed via rearing environment (rural vs. urban or suburban), pet ownership, pet care, and in one case 2 weeks of periodic observation of pets experiencing biological change (Herrmann et al., 2013). It is tempting to generalize across these direct experience of nature variables to argue that opportunities to closely observe nonhuman living things is the common thread and that the additional experiential variables of hands-on pet care and

Table 1. Summary of studies examining biological knowledge in children.

Author	*Participant Ages*	*Direct Experience Variable*	*Biological Knowledge Assessed; Task Details*	*Key Findings*
Prokop et al. (2008)	6–15 years	Pet ownership, number of pets	Internal structures; free drawing	Children who raised pets had significantly greater knowledge of the internal structure of a variety of animal species compared to children who did not own a pet ($p < .05$). This knowledge increased significantly for children who owned more than two pets ($ps < .001$) compared to children who only raised one pet and children who did not own a pet at all.
Prokop et al. (2007)	6–16 years	Pet ownership and personal observations of pets	Internal structures; free drawing	The researchers found a general increase in knowledge related to animal organ systems in children who had experience rearing a variety of animal species. This increase in knowledge fell just outside of significance at $p = .06$.
Myers et al. (2004)	4–14 years	Pet ownership, number of pets, care for pets	Understanding animal needs; interview	The number of pets owned was significantly correlated with knowledge of ecological and conservation needs. The more pets a child owned, the greater his or her understanding of ecological and conservation needs. Being responsible for caring for their pets led to higher scores on all need subsets, with ecological, conservation, and total needs scores being significantly higher for children with responsibility for pets compared to children without such responsibility.
Inagaki (1990)	5–6 years	Raising a pet	Factual knowledge and biological reasoning; interview	The goldfish-raising group possessed more factual knowledge about goldfish, made more reasonable predictions about goldfish ($p < .01$), and made significantly more goldfish analogies when predicting about frogs ($p < .05$).
Prokop & Tunnicliffe (2010)	10–15 years	Pet ownership, number of pets	Knowledge of and attitudes toward animals; interview	When age was controlled, results showed that pet owners generally displayed better knowledge of each animal than those without pets: Study 1 (pests; $\eta_p^2 = .04$), Study 2 (predators; $\eta_p^2 = .04$), Study 3 (disease; $\eta_p^2 = .01$). Animal knowledge increased as the number of pets increased, with children who owned three or more pets displaying the greatest degree of knowledge for all but one animal.
Toyama et al. (1997)	5–6 years	Caring for animals	Biological knowledge of care procedures; direct observation and interview	Carers showed greater understanding of the biological meaning of species-specific care behavior and made more accurate predictions than noncarers. Carers were also more likely to provide biological explanations for nonbiological procedures than their noncarer peers.

(*Continued*)

Table 1. (Continued).

Author	*Participant Ages*	*Direct Experience Variable*	*Biological Knowledge Assessed; Task Details*	*Key Findings*
Melson & Fogel (1989)	4–8 years	Pet ownership	Ideas about animal young and how animals care for their young	Children with pets provided more descriptions about animal care behaviors than children without pets.
Geerdts et al. (2015a)	3–5 years	Pet ownership	Biological reasoning; property projection	Both younger and older children with pets showed less anthropocentric reasoning than their non-pet-owning counterparts. There was a significant interaction between pet owners and whether they were attributing psychological or biological properties (η_p^2 = .57). Younger and older children with pets were both more likely to attribute biological properties to pets than children without pets (3-year-olds: d = 0.98, 5-year-olds: d = 0.78).
Williams & Smith (2006)	4–14 years	Urban vs. rural, pet ownership, pets that had given birth	Inheritance knowledge; structured interview	Pet owners made significantly more sophisticated judgements and explanations about parents' equal contribution of inherited traits (judgements: η_p^2 = .03, explanations: η_p^2 = .05) and more sophisticated explanations of social and biological aspects of inheritance (η_p^2 = .04) when explaining the inheritance of traits not present in parents (red hair with black-haired parents; η_p^2 = .03). When the researchers controlled for age they found that children gave more sophisticated explanations about the causal mechanisms of inheritance (η_p^2 = .04). Children whose pet had given birth showed a better understanding of the contribution of both parents to inherited traits (η_p^2 = .10). When age was controlled, results showed that rural children displayed a better understanding of parents' contribution of inherited traits and provided more sophisticated explanations of the social and biological aspects of inheritance (η_p^2 = .04) than their urban and suburban peers.
Panagiotaki & Nobes (2014)	4–7 years	Rural vs. urban	Organ function knowledge; structured interview	British children possessed significantly greater understanding of the location and functions of organs (η_p^2 = .20). Pakistani children made significantly more life appeals for the need for organs than British children (η_p^2 = .13).
Coley et al. (2005)	Kindergarten–Grade 6	Rural vs. urban (population density)	Ecological reasoning; property projection justifications	A significant difference was detected between rural and urban children's use of ecological reasoning when making inferences about disease. Population density was a significant predictor of ecological reasoning (along with property type and item type; R^2_{adj} = .308), with ecological reasoning decreasing as population density increased. A significant difference in ecological reasoning was also detected between local and exotic species, $F(1, 115) = 6.77$, $MSE = 11.30$, $p = .010$, with significantly more ecological references being made for local species than exotic species.

(Continued)

Table 1. (Continued).

Author	*Participant Ages*	*Direct Experience Variable*	*Biological Knowledge Assessed; Task Details*	*Key Findings*
Coley (2012)	6–10 years	Rural vs. urban (population density), experiences and activities	Ecological reasoning; property projection justifications	Higher nature experience scores were associated with living in more rural environments. Age, population density (p = .032), and nature experiences (p = .019) all significantly predicted ecological reasoning (R^2_{adj} = .308).
Herrmann et al. (2013)	3–4 years	Two-week exposure to animals in a classroom setting	Knowledge of natural biological changes across the life span	Postobservation children showed a significant increase in biological reasoning (η_p^2 = .10). They made significantly more endorsement of change after the experience (η_p^2 = .20) and showed better biological reasoning for their endorsements (η_p^2 = .88).
Medin et al. (2010)	4–10 years	Rural vs. urban	Biological reasoning; property projection	A significant interaction between community type and base species was detected for 4- to 5-year-olds, with rural children more likely than urban children to generalize from a dog base to nonhuman animal targets. However, both groups performed comparably when the base was a human. No significant differences between rural and urban children were found in relation to asymmetries favoring humans as the inductive base.
Ross et al. (2003)	6–10 years	Rural vs. urban	Ecological reasoning; property projection	Ecological reasoning was highest in older rural children.
Rusca & Tonucci (1992)	6–10 years	Rural vs. urban	Types of animals known; free list	Rural children mentioned farmyard animals twice as many times as urban children and listed many more birds (30%) than urban children (14.8%).
Tarlowski (2006)	4–5 years	Rural vs. urban, parental transmission (layperson vs. biological expert)	Biological reasoning; property projection	Rural children made fewer projections than urban children (p < .001). Rural children were more aware of the division between animate and inanimate groups than urban children. Experts' children made fewer projections than laypeople's children (p < .05), making sharper distinctions at the proper animate–inanimate category boundary than their layperson peers but appearing to use the category "animal" when reasoning. Rural expert children made significantly fewer projections than either urban group, suggesting an interaction effect of experience and parental transmission.

rural living may not further contribute to the development of biological knowledge. However, the diverse operationalization of biological knowledge makes it difficult to compare results. The rearing environment studies provide evidence suggesting that a rural upbringing fosters both a broad understanding of how living things relate to one another in the biological domain (Ross et al., 2003) and an understanding of specific biological phenomena such as the types of traits that are inherited from parents (Williams & Smith, 2006). The pet studies, however, tend to focus more tightly on children's knowledge and understanding of animal- and, in many cases, pet-relevant facts and phenomena. In future work, it will be important to be more systematic in identifying which elements of biological understanding are more or less influenced by direct experience and to design studies that isolate different types of direct experience of nature. For instance, it may be that the types of natural events that rural life makes accessible, such as witnessing animal birth, are primarily important for understanding the biology of inheritance and birth, but that different types of experience engender the relatively sophisticated biological reasoning patterns that are characteristic of rural children.

The underlying assumption in this literature is that direct experience of nature causes the development of biological knowledge. This is premature, however, as the research to date is almost exclusively correlational. The one intervention study done so far does suggest that direct experience of the natural world drives change in biological understanding (Herrmann et al., 2013). This is a good start, but the underlying mechanism is still unclear: Does exposure to nature give children access to facts about the natural world via observation? Or is it that such direct experience primes children to attend more closely to biological phenomena and related information presented by parents, peers, school, and the media? Or could it be that direct experience with nature motivates children to actively seek out new information about biology? Further studies exploring the roles of culture and parental transmission will help to answer these questions.

There is a general consensus among researchers in both psychology and education that parents play in integral role in helping to structure and develop their child's understanding of the world (Geerdts, Van de Walle, & LoBue, 2015b; Matalon, 2000; McNeil, 1982; Tunnicliffe & Reiss, 2000). It is widely accepted that parental input and daily experiences are important in a child's emerging understanding of biological and natural phenomena, including ecology, plants (Tunnicliffe & Reiss, 2000), animals (Geerdts et al., 2015b; Shepardson, 1997), the human body, and death (Matalon, 2000; McNeil, 1982).

However, to date few studies have examined the effects of parental input on children's understanding of biology and how this may occur in informal settings. Most research in this area has been conducted in informal learning environments such as zoos, aquariums, museums, and science centers (Geerdts et al., 2015b). However, further research needs to explore the potential interaction and influence of parental factors on children's biological knowledge, pet ownership, and contact with nature. It is possible that parental factors such as an interest in animals, feelings toward nature, and work type (e.g., biology) may influence a child's direct experience with nature and how parents communicate with their child about nature (see, e.g., Tarlowski, 2006).

The effects of direct experience on children's biological knowledge have been explored by both educational researchers and developmental psychologists; however, very little communication or collaboration between these disciplines has occurred. To date, most researchers within developmental psychology have failed to cite relevant educational research, with the exception of a few more recent studies (e.g., Geerdts et al., 2015a; Herrmann et al., 2013), preferring to concentrate on developments within their own discipline. Greater communication and collaboration between the various disciplines involved would undoubtedly benefit further research.

In addition to the outstanding basic research questions, numerous practical issues arise from the findings reviewed here. It is not yet clear whether the positive association between children's biological knowledge and a rural upbringing can be reduced to pet ownership and care because this is confounded by the fact that more rural than urban households have pets (Leslie, Meek, Kawash, & McKeown, 1994; Murray, Browne, Roberts, Whitmarsh, & Gruffydd-Jones, 2010). It will be important to dissociate these two effects in future work, because if pet ownership is responsible for rural children's superior biological knowledge, then this is a readily accessible avenue to intervention for nonrural children. Another question is whether the pet ownership effect would be similar for different types of animals; Inagaki's work indicating relatively advanced biological understanding in both mammal- and goldfish-owning children suggests that keeping pet goldfish may be a relatively easy, inexpensive means of increasing urban children's direct experience of nature (Inagaki, 2002, cited in Inagaki & Hatano, 2006). Also, the studies suggest that pets at home and in school are both effective at facilitating children's biological understanding, but a direct comparison has not been carried out. Again, from a practical perspective this is an important question for designing optimal interventions for modern children who spend less time outdoors than previous generations (Malone, 2007) and therefore may miss out on the cognitive benefits of directly experiencing the natural world.

References

Atran, S., Medin, D., & Ross, N. (2004). Evolution and devolution of knowledge: A tale of two biologies. *Journal of the Royal Anthropological Institute, 10*(2), 395–420. doi:10.1111/j.1467-9655.2004.00195.x

Australian Council for Educational Research. (2002). *Rural and urban differences in Australian education* (LSAY Briefing Report No. 5). Retrieved from http://research.acer.edu.au/lsay_briefs/4

Balmford, A., Clegg, L., Coulson, T., & Taylor, J. (2002, March 29). Why conservationists should heed Pokémon. *Science, 295*(5564), 2367. doi:10.1126/science.295.5564.2367b

Carey, S. (1985). *Conceptual change in childhood.* Cambridge, MA: MIT Press.

Charles, C., & Wheeler, K. (2012). *Children and nature worldwide: An exploration of children's experiences of the outdoors and nature with associated risks and benefits.* Retrieved from the Children and Nature Network website: https://www.childrenandnature.org/wp-content/uploads/2015/04/CECCNNWorldwideResearch.pdf

Coley, J. (2012). Where the wild things are: Informal experience and ecological reasoning. *Child Development, 83*(3), 992–1006. doi:10.1111/j.1467-8624.2012.01751.x

Coley, J. D., Vitkin, A. Z., Seaton, C. E., & Yopchick, J. E. (2005). *Effects of experience on relational inferences on children: The case of folk biology.* Retrieved from http://csjarchive.cogsci.rpi.edu/Proceedings/2005/index.htm

Friedmann, E., & Son, H. (2009). The human-companion animal bond: How humans benefit. *Veterinary Clinics: Small Animal Practice, 39*(2), 293–326. doi:10.1016/j.cvsm.2008.10.015

Geerdts, M. S., Van de Walle, G. A., & LoBue, V. (2015a). Daily animal exposure and children's biological concepts. *Journal of Experimental Child Psychology, 130*, 132–146. doi:10.1016/j.jecp.2014.10.001

Geerdts, M. S., Van de Walle, G. A., & LoBue, V. (2015b). Parent–child conversations about animals in informal learning environments. *Visitor Studies, 18*(1), 39–63. doi:10.1080/10645578.2015.1016366

Hatano, G., & Inagaki, K. (2013). *Young children's thinking about biological world.* New York, NY: Psychology Press.

Herrmann, P. A., French, J. A., DeHart, G. B., & Rosengren, K. S. (2013). Essentialist reasoning and knowledge effects on biological reasoning in young children. *Merrill-Palmer Quarterly, 59*(2), 198–220. doi:10.1353/mpq.2013.0008

Inagaki, K. (1990). Young children's everyday biology as the basis for learning school biology. *Bulletin of Faculty of Education, Chiba University, 38*, 177–184.

Inagaki, K., & Hatano, G. (2006). Young children's conception of the biological world. *Current Directions in Psychological Science, 15*(4), 177–181. doi:10.1111/j.1467-8721.2006.00431.x

Leslie, B. E., Meek, A. H., Kawash, G. F., & McKeown, D. B. (1994). An epidemiological investigation of pet ownership in Ontario. *Canadian Veterinary Journal, 35*(4), 218–222.

Malone, K. (2007). The bubble-wrap generation: Children growing up in walled gardens. *Environmental Education Research, 13*(4), 513–527. doi:10.1080/13504620701581612

Matalon, T. H. (2000). *The relationship among children's conceptualization of death, parental communication about death, and parental death anxiety* (Doctoral dissertation). Available from ProQuest Dissertations and Theses Database. (UMI No. 9960950)

McNeil, J. N. (1982). Young mothers' communication about death with their children. *Death Education, 6*(4), 323–339. doi:10.1080/07481188308252139

Medin, D. L., Waxman, S. R., Woodring, J., & Washinawatok, K. (2010). Human-centeredness is not a universal feature of young children's reasoning: Culture and experience matter when reasoning about biological entities. *Cognitive Development, 25*(3), 197–207. doi:10.1016/j.cogdev.2010.02.001

Melson, G. F., & Fogel, A. (1989). Children's ideas about animal young and their care: A reassessment of gender differences in the development of nurturance. *Anthrozoos, 2*(4), 265. doi:10.2752/089279389787057920

Muldoon, J., Williams, J., Lawrence, A., Lakestani, N., & Currie, C. (2009). *Promoting a "duty of care" towards animals among children and young people: A literature review and findings from initial research to inform the development of interventions*. Retrieved from the University of Edinburgh, Child and Adolescent Health Research Unit, website: http://www.cahru.org/content/04-publications/04-reports/defra-yr-1-report-sept-09.pdf

Murray, J. K., Browne, W. J., Roberts, M. A., Whitmarsh, A., & Gruffydd-Jones, T. J. (2010). Number and ownership profiles of cats and dogs in the UK. *Veterinary Record, 166*(6), 163–168. doi:10.1136/vr.b4712

Myers, O. E., Jr., Saunders, C. D., & Garrett, E. (2004). What do children think animals need? Developmental trends. *Environmental Education Research, 10*(4), 545–562. doi:10.1080/1350462042000291056

Panagiotaki, G., & Nobes, G. (2014). Cultural influences on children's understanding of the human body and the concept of life. *British Journal of Developmental Psychology, 32*(3), 276–290. doi:10.1111/bjdp.12039

Patrick, P., Havu-Nuutinen, S., Byrne, J., Sigurjonsdottir, H., Tunnicliffe, S. D., Oskarsdottir, G., & Carvalho, G. S. (2013). Students (ages 6, 10, and 15 years) in six countries knowledge of animals. *NorDiNa, 9*(1), 18–32.

Patrick, P., & Tunnicliffe, S. D. (2011). What plants and animals do early childhood and primary students' name? Where do they see them? *Journal of Science Education Technology, 20*(5), 630–642. doi:10.1007/s10956-011-9290-7

Planet Ark. (2011). *Climbing trees: Getting Aussie kids back outdoors* (Research report). Retrieved from http://treeday.planetark.org/research/

Planet Ark. (2013). *Missing trees: The inside story of an outdoor nation* (Research report). Retrieved from http://treeday.planetark.org/research/

Prokop, P., Prokop, M., & Tunnicliffe, S. D. (2008). Effects of keeping animals as pets on children's concepts of vertebrates and invertebrates. *International Journal of Science Education, 30*(4), 431–449. doi:10.1080/09500690701206686

Prokop, P., Prokop, M., Tunnicliffe, S. D., & Diran, C. (2007). Children's ideas of animals' internal structures. *Journal of Biological Education, 41*(2), 62–67. doi:10.1080/00219266.2007.9656064

Prokop, P., & Tunnicliffe, S. D. (2010). Effects of having pets at home on children's attitudes toward popular and unpopular animals. *Anthrozoos, 23*(1), 21–35. doi:10.2752/175303710X12627079939107

Ross, N., Medin, D., Coley, J. D., & Atran, S. (2003). Cultural and experiential differences in the development of folkbiological induction. *Cognitive Development, 18*(1), 25–47. doi:10.1016/S0885-2014(02)00142-9

Rusca, G., & Tonucci, F. (1992). Development of the concepts of living and animal in the child. *European Journal of Psychology of Education, 7*(2), 151–176. doi: 10.2752/089279303786992134

Shafto, P., & Coley, J. D. (2003). Development of categorization and reasoning in the natural world: Novices to experts, naïve similarity to ecological knowledge. *Journal of Experimental Psychology: Learning, Memory, and Cognition, 29*(4), 641–649. doi:10.1037/0278-7393.29.4.641

Shepardson, D. P. (1997). Of butterflies and beetles: First graders' ways of seeing and talking about insect life cycles. *Journal of Research in Science Teaching, 34*(9), 873–889. doi:10.1002/(SICI)1098-2736(199711)34:9<873::AID-TEA3><873::AID-TEA3>3.0.CO;2-R

Tarlowski, A. (2006). If it's an animal it has axons: Experience and culture in preschool children's reasoning about animates. *Cognitive Development, 21*(3), 249–265. doi:10.1016/j.cogdev.2006.02.001

Toyama, N., Lee, Y. M., & Muto, T. (1997). Japanese preschoolers' understanding of biological concepts related to procedures for animal care. *Early Childhood Research Quarterly, 12*(3), 347–360. doi:10.1016/S0885-2006(97)90008-9

Tunnicliffe, S. D., & Reiss, M. J. (1999). Building a model of the environment: How do children see animals? *Journal of Biological Education, 33*(3), 142–148. doi:10.1080/00219266.1999.9655654

Tunnicliffe, S. D., & Reiss, M. J. (2000). Building a model of the environment: How do children see plants? *Journal of Biological Education, 34*(4), 172–177. doi:10.1080/00219266.2000.9655714

Uitto, A., Juuti, K., Lavonen, J., & Meisalo, V. (2006). Students' interest in biology and their out-of-school experiences. *Journal of Biological Education, 40*(3), 124–129. doi:10.1080/00219266.2006.9656029

Waxman, S., & Medin, D. (2007). Experience and cultural models matter: Placing firm limits on childhood anthropocentrism. *Human Development, 50*(1), 23–30. doi:10.1159/000097681

Waxman, S., Medin, D., & Ross, N. (2007). Folkbiological reasoning from a cross-cultural developmental perspective: Early essentialist notions are shaped by cultural beliefs. *Developmental Psychology, 43*(2), 294–308. doi:10.1037/0012-1649.43.2.294

Williams, J. M., & Smith, L. A. (2006). Social and experiential influences on the development of inheritance concepts. *International Journal of Behavioral Development, 30*(2), 148–157. doi:10.1177/0165025406063630

Zhang, W., Goodale, E., & Chen, J. (2014). How contact with nature affects children's biophilia, biophobia and conservation attitude in China. *Biological Conservation, 177*, 109–116. doi:10.1016/j.biocon.2014.06.011

Young Chinese Children's Justifications of Plants as Living Things

Ying Tao

ABSTRACT

Research Findings: The purpose of this study was to explore how Chinese preschool children categorize plants into either living or nonliving things. The research was framed within the interpretive paradigm and was designed as a descriptive, cross-sectional study. Participants were children 4 to 6 years of age from 3 kindergartens in Jiangsu Province in eastern China ($n = 90$). In-depth interviews were carried out to explore the justifications children used to classify trees, flowers, and grass as living or nonliving things. Six types of justifications were identified among the participating children. Cross-sectional comparisons of children's justifications were made both quantitatively and qualitatively. A developmental pathway of children's justifications of plants as living things was proposed. *Practice or Policy*: The results indicated that children's justifications of plants as living things undergo gradual change from 4 to 6 years of age. Moreover, the development of these justifications may be influenced by the accumulation of biological knowledge. Implications for early science instruction on young children's understanding of plants as living things are discussed.

Plants are a fundamental kingdom of living things that play a vital role in the complexities of all ecosystems on Earth. It seems obvious, therefore, that "a basic understanding of how plants live and grow is essential as we plan for the uncertainties in our future" and that "early childhood is the ideal time to begin that training" (Archer, 2014, p. 575). Children are able and ready to learn about plants in their natural surroundings from the earliest ages (Ministry of Education [China], 2012; National Science Teachers Association, 2014). Children most often construct understandings of plants through direct, concrete experiences, for example, by observing, touching, smelling, and feeling their texture (Tunnicliffe, 2001; Tunnicliffe & Reiss, 2000). Compared with the extensive research on the early emergence of the concept of animal (e.g., Carey, 1985; Erickson, Keil, & Lockhart, 2010; Keil, 1989; Piaget, 1929; Springer, 1999), the study of children's conceptions of plants has only occasionally been the focus of research attention (Gatt, Tunnicliffe, Borg, & Lautier, 2007; Nguyen & Gelman, 2002; Tunnicliffe, 2001).

Researchers have found that many preschoolers know that plants possess specific characteristics such as leaves, roots, and stems and have basic needs, including air, water, and sunlight (Barman, Stein, McNair, & Barman, 2006; Villarroel & Infante, 2014). Moreover, a number of previous studies demonstrated that many preschoolers recognize the commonalities between plants and animals and how they both differ from nonliving things: in the capacity to grow (Hatano et al., 1993; Inagaki & Hatano, 1996; Stavy & Wax, 1989), to regrow and heal in the face of injury (Backscheider, Shatz, & Gelman, 1993; Hatano et al., 1993), to reproduce (Springer & Keil, 1991), to decompose (Springer, Ngyuen, & Samaniego, 1996), and to die (Hatano et al., 1993; Inagaki & Hatano, 1996; Nguyen & Gelman, 2002).

However, children's knowledge of plants' perceptual characteristics and their recognition of plants' biological processes do not usually lead them to conclude that plants, like animals, are living things (e.g., Opfer & Siegler, 2004; Stavy & Wax, 1989; Villarroel & Infante, 2014). A considerable body of research indicates that young children have difficulty attributing life status to plants (e.g., Carey, 1985; Hatano et al., 1993; Martínez-Losada, García-Barros, & Garrido, 2014; Piaget, 1929; Richards & Siegler, 1984; Venville, 2004). Research has also found that children tend to attribute life status first to human beings, then to animals, and finally to plants (Carey, 1985; Hickling & Gelman, 1995; Inagaki & Hatano, 2002; Leddon, Waxman, & Medin, 2009; Nguyen & Gelman, 2002; Richards & Siegler, 1986; Stavy & Wax, 1989; Yorek, Sahin, & Aydin, 2009).

One explanation for the difficulty children have attributing life status to plants is that plants' visual properties make it difficult for children to classify them as living things. Unlike animals, plants have no obvious organs and can be perceptually very different from one another, for instance, trees and mosses (Nguyen & Gelman, 2002; Richards & Siegler, 1986). Another explanation emphasizes analogical mapping with human beings (Inagaki & Hatano, 1996, 2002). That is, preschoolers initially organize their understanding of life status around people's characteristics: The more closely an entity resembles people, the more likely it is that the entity will be judged to have life status. Young preschoolers (3- and 4-year-olds) may initially map the living/nonliving distinction onto many features that are characteristic of prototypic classes of living things, people, and other animals. Among the features that young preschoolers typically view as characteristic of living things are the presence of eyes, a mouth, and moving legs (S. A. Gelman & Opfer, 2007; Martínez-Losada et al., 2014; Richards & Siegler, 1986); or the capacity to do different things, such as eat and drink (Lucas, Linke, & Sedgwick, 1979). All of this is consistent with the idea that young children use analogical reasoning, meaning that they extend human qualities and behavior to other living beings (Carey, 1985).

Another explanation for children's difficulty in integrating plants into the biological domain is that young children tend to use the criterion of movement to justify objects as living or nonliving. Children often expect things that are alive to move (Margett & Witherington, 2011; Solomon & Zaitchik, 2011); thus, movement occupies a central position in children's construction of a biological concept of life (Opfer & Siegler, 2004; Zogza & Papamichael, 2000). Plants do not spontaneously move and thus do not naturally attract children's attention in the same way that animals do (Tunnicliffe, 2001). Therefore, in order to categorize plants as living organisms, children need to be able to overcome the cognitive obstacle of a plant's apparent motionlessness and acknowledge that plants do not have to move to be alive (Brulé, Labrell, Megalakaki, Fouquet, & Caillies, 2014).

Although these studies have revealed young children's knowledge about plants and their difficulty in attributing life status to plants, the research has not explored in detail when and how children gradually categorize plants as living organisms during the preschool years or what type of experiences lead children to conclude that plants are alive (Brulé et al., 2014; Opfer & Siegler, 2004). In addition, the limited number of published studies on preschool children's knowledge of plants in the cultural context of China means that Chinese preschool children's justifications of plants as living or nonliving things remain largely undocumented.

Method

Purpose and Research Questions

Given the gaps in the literature identified in the previous section, the purpose of the current study was to explore Chinese preschool children's intuitive understanding of plants as living or nonliving things by investigating whether they can recognize trees, flowers, and grass as living things and ascertaining the justifications that they use for their views. Moreover, the purpose of the study was to understand how young Chinese children's justifications for plants as living or nonliving things develop during the preschool years.

More specifically, the research questions were as follows:

(1) Do young Chinese children classify trees, flowers, and grass as living or nonliving things?
(2) What types of justifications do young Chinese children use to classify trees, flowers, and grass as living or nonliving things?
(3) How do young Chinese children's justifications of the life status of plants develop during the preschool years?

Research Design

The research presented in this article was a cross-sectional study conducted in three kindergartens in China. Consistent with the research questions, which focus on children's in-depth understanding of plants as living things, this research was conducted within an interpretive frame (Cohen, Manion, & Morrison, 2000). The data presented in this article were generated through detailed qualitative interviews with participating children. In order to address Research Questions 1 and 2 about whether young children recognize plants as living things and the criteria they use in these classifications, I designed the research to be descriptive through the use of in-depth interviews. Descriptive research "set[s] out to describe and interpret *what is,*" and the "researcher accounts for what has already occurred" (Cohen & Manion, 1989, p. 70, emphasis in the original). In order to address Research Question 3, about how children's justifications of plants as living things develop during the preschool years, the research design incorporated a cross-sectional element with data collected from 4-, 5-, and 6-year-old children. In education, "cross-sectional studies involve indirect measures of the nature and rate of changes in the physical and intellectual development of samples of children drawn from representative age levels" (Cohen et al., 2000, p. 175).

Participants

Participants were 90 young children (50 boys and 40 girls) from three public kindergartens in Jiangsu Province in eastern China. All kindergartens were known to the researcher through her professional associations. Children attending the kindergartens had spent time in structured approaches to language, social studies, and art but had not yet received any structured instruction on living things. Therefore, children from the three kindergartens were considered appropriate participants for the present study because it would be possible to reveal their untutored and intuitive ideas about plants.

In each kindergarten, 10 children were randomly selected from three age groups, namely, 4-, 5-, and 6-year-olds, as a number of previous studies have found this age span to be a key period in which important changes in children's understandings of plants take place (e.g., Backscheider et al., 1993; Brulé et al., 2014; Hickling & Gelman, 1995; Nguyen & Gelman, 2002). Thus, in total 30 children were selected from each age group. When the study was conducted, the mean age of the 4-year-old participants (17 boys and 13 girls) was 49.6 months (SD = 2.9), the mean age of the 5-year-old participants (16 boys and 14 girls) was 60.9 months (SD = 3.4), and the mean age of the 6-year-old participants (17 boys and 13 girls) was 73.1 months (SD = 3.5).

Data Collection

Research with the young children in the three kindergartens was not common prior to this study. Consequently, a pilot study with a small number of children indicated that when they were invited for individual interviews in the staff room or library, rather than their homeroom, they portrayed a level of nervousness because they were reluctant to respond my prompts and questions. However, when the children came in pairs, they seemed to be more relaxed and at ease expressing their ideas. Therefore, face-to-face interviews with pairs of children were used to probe their understanding of plants as living or nonliving things. The researcher established rapport with the children by ensuring that they understood that the researcher was interested in their ideas about plants, that the interview was not a test and would not be scored, and that no one would be told about their individual performance.

The interview protocol was based on the researcher's previous studies of elementary children's conceptual understandings of living things and nonliving things (Tao, Oliver, & Venville, 2012a, 2012b). The interviews were conducted in Mandarin. The Chinese translation for the question "Is X living?" is "*X you sheng ming ma?*" which literally means "Has X got life?" Three color pictures (16 × 15 cm) respectively displaying plane trees, daisies, and lawn grass were used to prompt the children to discuss whether they thought these items are living or not. These three types of plants are very common in Jiangsu Province and can be easily found in the environment surrounding the kindergartens. The children initially were asked whether they knew what the picture was and then were asked to point out the items that they thought are living. The children then were encouraged to give reasons for their responses, for example, "Why do you think X is living/not living?" "Could you please tell me more about that?" "How do you know that?" and so on. Each interview took around 20 min, and all interviews were audio recorded and fully transcribed.

Data Analysis

The data generated by the interviews were analyzed in four stages. In the first stage, percentages of children's correct responses to the questions about whether the trees, flowers, and grass are living were calculated. Scoring of the responses involved allocating 1 point for the correct answer for each item. Descriptive statistics were generated for each age group, interpreted, and compared. Also, one-way between-groups analysis of variance (ANOVA) was conducted to compare the scores for the three items among different age groups. In the second stage, types of justifications used by the participating children to determine the life status of trees, flowers, and grass were identified. In the third stage, the children's responses were coded in accordance with the types of justifications they provided. For example, when a child responded that "trees would feel pain when people cut them down," an anthropomorphic code was given. In some cases, two codes were attributed to one answer because children used various types of justifications to determine the life status of a single item. For example, when a child responded that "flowers have got petals, stalks, and stems and they can wither," both a perceptual code and a biological code were assigned. Graphs presenting types of justifications by the children's age were generated by computer and ANOVA was conducted so that cross-sectional comparisons could be made. In the final stage, each child's responses to all interview questions were considered holistically and the internal consistency of types of justifications used by each child to determine the life status of trees, flowers, and grass was examined. Children with similar responses to the interview questions were clustered into categories.

Excerpts from in-depth interviews were selected for inclusion here if they were representative of a certain type of justification. All children were assigned pseudonyms, and no actual names are used. Each child's age in months is provided, indicated by M. So "Niu Niu, 54 M" means that this child's pseudonym is "Niu Niu" and she was 54 months old at the time of the interviews.

Results

Children's Correct Responses by Age Group

The mean scores for the participating children's correct responses to the questions about whether trees, flowers, and grass are living are displayed in Table 1. In order to examine whether there were significant differences among the three age groups in relation to their correct responses to whether trees, flowers, and grass are living things, a one-way ANOVA was conducted for each item at a time. All three test results were statistically significant, indicating that the children's correct responses varied considerably between age groups. These findings were consistent for trees, flowers, and grass: trees, $F(2, 87) = 3.25$, $p < .05$, $\eta^2 = .069$; flowers, $F(2, 87) = 3.73$, $p < .05$, $\eta^2 = .079$; grass, $F(2, 87) = 7.23$, $p < .01$, $\eta^2 = .143$. Subsequent post hoc analysis with Bonferroni correction (using $\alpha = .05$) was also conducted for each item

Table 1. Mean Scores for Children's Responses to Each Item by Age Group.

	4-Year-Olds		5-Year-Olds		6-Year-Olds	
Item	M	SD	M	SD	M	SD
Trees	0.83	0.379	0.73	0.450	0.97	0.183
Flowers	0.83	0.379	0.67	0.479	0.93	0.254
Grass	0.77	0.430	0.60	0.563	1.00	0.000

at a time, and the results revealed statistically significant differences between 5-year-old children and 6-year-old children. Five-year-old children had significantly lower mean scores compared with 6-year-old children in terms of their correct responses to whether trees, flowers, and grass are living. There were no other significant differences between the other groups.

Types of Children's Justifications

The analysis of the participating children's responses and explanations to questions about why they thought trees, flowers, and grass are living or not living indicated six types of justifications: perceptual justification, anthropomorphic justification, motion justification, intact justification, functional justification, and biological justification.

Perceptual Justification

This type of justification was invoked when children referred to visible phenomena that they can see. In many cases this included the observable characteristics that trees and grass have, for example, leaves and stems; are green; and are planted in the soil or ground. For flowers, perceptual justifications often focused on blooms in spring and buds and petals. Quite a number of children mentioned that tree leaves and grass turn yellow in winter. For example:

Researcher: Do you think flowers are living?
Su Jin (48 M): Yes, flowers have petals and a green stem, and they smell lovely.
Researcher: Do you think trees are living?
Zi Xuan (53 M): Yes, trees have leaves and I see them every day.
Researcher: Do you think trees are living?
Li Xu (67 M): Yes, the tree leaves used to be green, and they turned yellow and orange in winter.

Anthropomorphic Justification

This type of justification refers to children ascribing human characteristics to organisms and often had a connection with children's own experiences. In many cases children used "feeling pain" to justify the life status of trees, flowers, and grass. These children tended to believe that trees, flowers, and grass would have feelings when people cut them down, pull them up, or pick them. Also, children mentioned that trees, flowers, and grass "drink water" as we do. Alternatively, children tended to deny the life status of trees, flowers, and grass by constantly comparing them with human beings. They affirmatively responded that living things should have "eyes," a "nose," and a "mouth"; should "go home at night"; and can "talk." The most common expression in children's responses was that "they are not like us." For example:

Researcher: Do you think trees are living?
Meng Meng (47 M): Yes, a tree would feel pain when people cut it down, because that's its body.
Researcher: What about flowers? Are flowers living?
Meng Meng: Yes, flowers are living because they would feel pain when we pick off their petals.
Researcher: Is grass living as well?
Meng Meng: Yes, grasses would feel pain if we pull them up from the soil.
Researcher: Do you think trees are living?

Zi Ming (54 M):	No, they come out from the ground, they are not like us.
Researcher:	Well, then how can things be living?
Zi Ming:	Things that don't come out from the ground, things that eat rice and grow taller like we do.
Researcher:	Do you think flowers and grass are living?
Zi Ming:	No, they all come out from the ground.

Motion Justification

This type of justification occurred when children referred to movement or the absence of movement when determining the life status of trees, flowers, and grass. The in-depth interviews revealed that many children believed that movement or mobility is an essential attribute of all living things. A number of children responded that trees, flowers, and grass are living because they "wave in the wind," they "bend down in a storm," and "flowers open," whereas others denied the life status of these plants because they perceived that they do not move by themselves. For example:

Researcher:	Do you think trees are living?
Niu Niu (54 M):	Yes, trees wave their leaves when the wind blows.
Researcher:	Do you think grass is living?
Duo Duo (68 M):	Yes, some grass can actually move. Mimosa closes its leaves when people touch it.
Researcher:	Do you think flowers are living?
Han Yan (78 M):	Yes, flowers can actually move. Their petals move slowly when flowers open.
Researcher:	Do you think trees are living?
Mo Han (61 M):	No, trees don't move at all. Tree leaves can wave sometimes, but the tree trunk doesn't move at all.

Intact Justification

This type of justification refers to children's belief that living things are those that never die and remain alive or intact. During the in-depth interviews the children using this type of justification indicated that trees are living because they are tough, robust, or will not be pulled over, whereas flowers and grass are not living because they could be damaged because of various factors. For example:

Researcher:	Do you think trees are living?
Le Le (43 M):	Yes, trees are very strong and they will not be pulled over by people.
Researcher:	What about flowers? Are flowers living?
Le Le:	No, flowers can be picked off by people.
Researcher:	Well, can you think of anything else that is living?
Le Le:	Apples and bananas. They are living before we eat them.

Functional Justification

This type of justification refers to children ascribing a function to the items. A few children attributed life status to trees, flowers, and grass because they believed that these organisms play important roles in our environment. For instance, "the world would be desert without trees" (Bo Bo, 70 M), trees "give us good oxygen and provide shelter from rain and wind" (Rui Rui, 66 M), and grass and flowers "bring beauty and fragrance to us" (Li Xu, 67 M).

Biological Justification

This type of justification consisted of four subjustifications—death, growth, having needs, and internal processes—which are explained in the following sections.

Death. In contrast to the previous intact justification, this justification refers to children's belief that living things will eventually die and cannot physically come back to life again. During the interviews, children sometimes explained that trees, flowers, and grass are living things because they will die because of natural or human factors. For example:

Researcher:	Are flowers living?
Rui Xi (48 M):	Yes, we should not pick flowers, because if we do, they will die.
Researcher:	Do you think trees are living?
Wei Lai (56 M):	Yes, they are. If people cut them down, they would not have life any more.
Researcher:	Do you think trees are living?
Heng Yu (70 M):	Yes, when trees get really old or really sick, they will break off.
Researcher:	What about flowers? Are flowers living?
Heng Yu:	Yes, flowers will eventually wither and die.
Researcher:	Is grass living as well?
Heng Yu:	Yes, grass will wither too.

Growth. The growth justification refers to children's explanations that trees, flowers, and grass are living things because they grow from "seeds" (Xiao Bao, 49 M) or "buds" (Ruo Xuan, 51 M) and "sprout" (Lin Lin, 60 M); they used to be "small and little" (Yun Han, 74 M); and they grow "bigger and taller" (Yu Chen, 72 M) as time goes by. Children's responses demonstrated an understanding of growth as an important attribute of living things. For example:

Researcher:	Is grass living?
Mo Mo (49 M):	Yes, it is. I saw people cut the grass with big scissors, but it grew back again.
Researcher:	Do you think trees are living?
Yan Ning (52 M):	Yes, trees have leaves and their leaves can grow.
Researcher:	Do you think trees are living?
Zi Ran (68 M):	Yes, trees used to be small seedlings and it takes a long time for trees to grow up to be what it looks like now.
Researcher:	Do you think flowers are living?
Chen Hao (68 M):	Yes. Flowers don't move, but they actually grow. They can grow into big blooms.

Having Needs. This justification refers to some children's belief that trees, flowers, and grass have various needs to sustain their life. During the interviews, children mentioned that trees, flowers, and grass are living things because they need "soil" (Yi Xin, 75 M), "water" (Shi Tou, 62 M), "rain" (Cheng Cheng, 66 M; Dou Dou, 64 M), "sunlight" (Xin Lin, 70 M), and "fertilizer" (Shi Tou, 62 M).

Internal Processes. Compared with the previous criteria, such as death and growth, which are visible and more obvious to young children, the ideas that trees "make their own food with sunlight, water and air" (Han Yu, 74 M), trees "breathe in carbon dioxide and exhale oxygen" (Zhi Yuan, 76 M), and flowers and grass "have roots that absorb and transfer water and nutrients" (Heng Heng, 70 M; Yue Yue, 74 M) are more complicated and abstract. However, a small number of children applied this type of justification in their explanation of why trees, flowers, and grass are living.

Quantitative Analysis of Justifications by Age Group

Children's mean scores for the six types of justifications by age group are displayed in Table 2. In order to examine whether there were significant differences among the three age groups in relation to the six types of justifications, a one-way ANOVA was conducted for each type of justification at a time. The results for perceptual justifications, motion justifications, and biological justifications were statistically significant,

Table 2. Children's Mean Scores for the Six Types of Justifications by Age Group.

	4-Year-Olds		5-Year-Olds		6-Year-Olds	
Types of Justification	M	SD	M	SD	M	SD
Perceptual	1.30	1.26	0.63	0.96	0.73	0.91
Anthropomorphic	0.60	1.16	0.23	0.73	0.10	0.40
Motion	0.40	0.62	0.93	1.31	0.17	0.38
Intact	0.37	0.96	0.10	0.55	0.10	0.55
Functional	0	0	0.10	0.55	0.30	0.65
Biological	1.03	1.03	1.37	1.19	1.93	1.11

indicating that these three justifications varied considerably between age groups. These findings were consistent for perceptual, motion, and biological justifications: perceptual, $F(2, 87) = 3.47, p < .05, \eta^2 = .074$; motion, $F(2, 87) = 4.63, p < .01, \eta^2 = .124$; biological, $F(2, 87) = 5.01, p < .01, \eta^2 = .103$. Subsequent post hoc analysis with Bonferroni correction (using $\alpha = .05$) was also conducted separately for each of the above justifications. The results revealed that the participating 4-year-old children ($M = 1.30$, $SD = 1.26$) had a significantly higher mean score for perceptual justifications than the 6-year-old children ($M = 0.73$, $SD = 0.91$); the participating 5-year-old children ($M = 0.93$, $SD = 1.31$) had a significantly higher mean score for motion justifications than the 6-year-old children ($M = 0.17$, $SD = 0.38$); and the participating 6-year-old children ($M = 1.93$, $SD = 1.11$) had a significantly higher mean score for biological justifications than the 4-year-old children ($M = 1.03$, $SD = 1.03$). Types of justifications by age group are displayed in Figure 1.

As displayed in Figure 1, 4-year-old children tended to rely on perceptual justifications to determine the life status of trees, flowers, and grass, 6-year-old children were more likely to use biological justifications, whereas the 5-year-olds seemed to rely on motion cues significantly more than the 6-year-olds. The 5-year-old children's preference for motion justifications is similar to results obtained in other studies conducted with preschool children and also confirms that motion or movement plays an important role in children's construction of a biological concept of life (Brulé et al., 2014; Yorek et al., 2009; Zogza & Papamichael, 2000).

The results also provide explanations for the previous finding that the children's correct responses to whether trees, flowers, and grass are living varied considerably among age groups (see Table 1). Aided by perceptual information that was most likely gained through daily observation and discussions with family members and friends, most 4-year-old children tended to believe that trees,

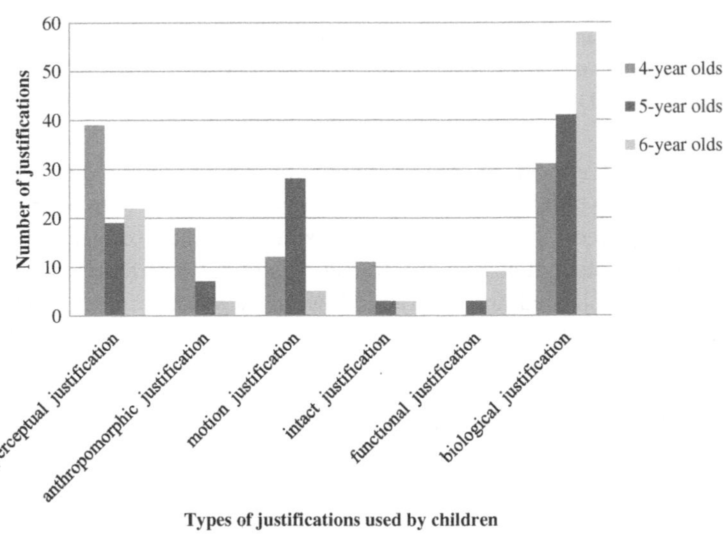

Figure 1. Types of justifications used by children by age group.

flowers, and grass are living. Because of their perception of plants' immobility, a number of 5-year-old children denied the life status of the three items, thus resulting in lower mean scores for correct responses for this age group. The findings indicate that as a result of developing comprehension of plants' inner properties, nearly all 6-year-old children responded that trees, flowers, and grass are living things. Biological justifications were most frequent among the 4-year-old children, which indicates that at that age, children were able to include plants, such as trees, flowers, and grass, in the biological domain.

There were no statistically significant differences among the three age groups in mean scores for anthropomorphic justifications (p = .057), intact justifications (p = .253), or functional justifications (p = .06). However, the descriptive analyses (see Table 2) showed that anthropomorphic justifications and intact justifications tended to decrease with age, whereas functional justifications tended to increase with age.

Qualitative Analysis of Children's Justifications

A closer examination of each child's responses to all interview questions across trees, flowers, and grass revealed that some children used a single justification all the way through, whereas others used multiple types of justifications. According to the internal consistency of the justifications used, six categories of responses were found among the participating young Chinese children. These categories are shown in Table 3.

Seven 4-year-old children and nine 5-year-old children used either an anthropomorphic justification, intact justification, or motion justification as their single justification across trees, flowers, and grass, revealing that these children relied on animistic or anthropocentric thinking (see Table 3, Category A). Children in this category either responded that trees, flowers, and grass are living because they feel pain when we cut them (Xuan Xi, 54 M) or argued that none are living because they do not look like us (Zi Ming, 54 M) or they do not move (Mo Han, 61 M). This category was not detected among the 6-year-old children.

A number of children tried to reconcile these animistic or anthropocentric criteria with their perceptual knowledge and/or biological knowledge, thus using various combinations of justifications. Unlike the previous category, in which children used the same justification across three types of plants, children clustered in Categories B, C, and D tended to use different types of justifications for different plants. Also, these children tended to provide contradictory answers to all interview questions. In Category B, for example, one 6-year-old child used a functional justification for trees, a perceptual justification for grass, and a motion justification for flowers (Su Ya, 76 M). In Category C, children tended to use a biological justification for one plant and a motion justification for the others, or vice versa. For example, one 4-year-old child and one 6-year-old child said that flowers and grass are living because they grow or die, but trees are not living because they do not move (Yu Xin, 46 M; Bo Ning, 72 M). In Category D, for example, one 4-year-old child who initially denied the life status of grass because grass does not have eyes, a nose, or a mouth did not use the same rule for flowers. She argued that flowers are living because they need water even if they do not have eyes, a nose, or a mouth (Ming Wei, 53 M).

Nine 4-year-old children, four 5-year-old children, and two 6-year-old children fell into Category E, in which they applied a perceptual justification to one plant and a biological justification to the others, or vice versa. Compared with all previous categories, anthropomorphic, intact, motion, and functional justifications were not detected among these children. A few more 4-year-old children were clustered into this category, revealing the younger children's tendency to rely on perceptual experiences when determining the life status of plants.

Six 4-year-old children, eleven 5-year-old children, and fifteen 6-year-old children fell into Category F, in which only biological justifications were used across trees, flowers, and grass. These children's consistent reference to the biological properties of plants indicates a self-regulating, coherent, and intuitive framework of biology (Venville, 2004). This is an interesting finding and

Table 3. Six Categories of Children's Use of Justification(s) Across Trees, Flowers, and Grass.

Category	*Description*	*Example(s) of Children's Responses*	*4-Year-Olds* (n = *30)*	*5-Year-Olds* (n = *30)*	*6-Year-Olds* (n = *30)*
A	Children consistently used one of the following justifications only: A/M/I; trees, flowers, and grass are not always living	"Trees, flowers, and grass are not living, because they don't talk or walk as we do and there is nothing in them." [A] "Trees, flowers, and grass are not living because they don't move." [M] "Trees and grass are living because they don't wither; flowers are not living because they wither." [I]	7 [6A, 1I]	9 [1I, 1A, 7 M]	0
B	Children combined P + M/I/F; trees, flowers, and grass are not always living	"Trees are living because they are beautiful." [F] "Flowers are not living because they don't move at all." [M] "Grass is living because it is planted in the ground." [P]	2 [P + I + M]	0	3 [P + F + A] [P + F]
C	Children combined B + A/M/I/F; trees, flowers, and grass are not always living	"Trees and flowers are living, because they will die when we cut them off." [death] "Grass is not living, because it doesn't feel pain or have any blood." [A]	3 [B + M]	4 [B + M] [B + A]	6 [B + M] [B + F] [B + I]
D	Children combined P + B + A/I/M/F; trees, flowers, and grass are not always living	"Trees are living because they have branches and leaves." [P] "Flowers are living, because they need water." [having needs] "Grass is not living, because it doesn't have eyes, mouth or a nose." [A]	3 [P + B + A] [P + B + I]	2 [P + B + M] [P + B + F]	4 [P + B + M] [P + B + F]
E	Children combined P + B; trees, flowers, and grass are always living	"Trees are living, because they have leaves." [P] "Flowers are living because they need water." [having needs] "Grass is living, because it turns yellow in winter." [P]	9	4	2
F	Children used B only; trees, flowers, and grass are always living	"Trees are living because they will die when we cut them off." [death] "Flowers are living because they grow from little buds." [growth] "Grass is living, because it grows from the soil." [growth]	6 [3 death] [2 growth] [1 having needs]	11 [4 death] [5 growth] [2 having needs]	15 [4 death] [5 growth] [3 having needs]; [3 internal processes]

Note. A = anthropomorphic justification; M = motion justification; I = intact justification; P = perceptual justification; F = functional justification; B = biological justification.

contrasts with other studies that have found that young children often think that plants are not living or children do not recognize plants as living until the age of 6 or even later (e.g., Gatt et al., 2007; Leddon et al., 2009; Opfer & Siegler, 2004). According to Richards and Siegler (1986), children's successful categorization of plants as living organisms requires not only the acknowledgment that they do not move but also gradual comprehension of plants' inner biological properties, such as growth, death, and having needs. During the in-depth interviews, a number of 6-year-old children explicitly responded that trees, flowers, and grass do not move but added comments such as "They've got life within their bodies" (Ruo Shui, 75 M), "They grow" (Zi Yu, 78 M), "Their leaves grow" (Chen Bo, 72 M), and "Their roots have got life" (Yue Yue, 74 M). These children's responses demonstrate the understanding that plants do not need to be perceived as able to move freely to be alive; moreover, children were more likely to indicate an understanding of the connection between life and the biological properties of plants. A closer examination of children's biological justifications indicated that the most frequently mentioned subjustifications were death, growth, and having needs. The internal processes justification, a much more complex justification, such as "Trees make their own food through sunlight and air," "Trees take in carbon dioxide and exhale oxygen," and "Roots have something within them to transfer nutrients from the soil," was provided by a small number of 6-year-old children. These justifications were not detected among the younger children.

Discussion

The aim of this cross-sectional study was to investigate whether young Chinese children classify trees, flowers, and grass as living things; what types of justifications the children tend to use in order to determine the life status of these three items; and how their justifications develop during the preschool years. The data presented in this study show that participating children between the ages of 4 and 6 years used a wide variety of justifications during the interviews when being probed about why they think trees, flowers, and grass are living or not living. These justifications were classified into groups: perceptual justification, anthropomorphic justification, motion justification, intact justification, functional justification, and biological justification. Two main trends were evident in the data with regard to the development of the participating children's understanding of plants as living things: Children's justifications of plants as living things undergo gradual changes from 4 to 6 years of age, and the development of children's justifications may be related to the accumulation of biological knowledge.

The Developmental Pathway of Children's Justifications

Figure 2 presents a diagrammatic representation of a proposed pathway in terms of how children's justifications of plants as living things develop during the preschool years. As the figure shows, there are qualitative changes in children's understandings of plants as living things from an initial anthropocentric and animistic framework to a more biologically oriented framework.

In Stage 1, children associate life with anthropomorphic and animistic characteristics; that is, the more closely an entity resembles people, the more likely it is that the entity will be judged to have life status. This could be accounted for by young children using analogical reasoning and extending human qualities and behavior to other unfamiliar living beings (Carey, 1985; Inagaki & Hatano, 2002, 2006). According to Inagaki and Hatano (2006), young children tend to draw analogically on their knowledge about humans and make "educated guesses" (p. 178) when attributing properties to less familiar living entities. Children's reference to human characteristics indicates that they possess

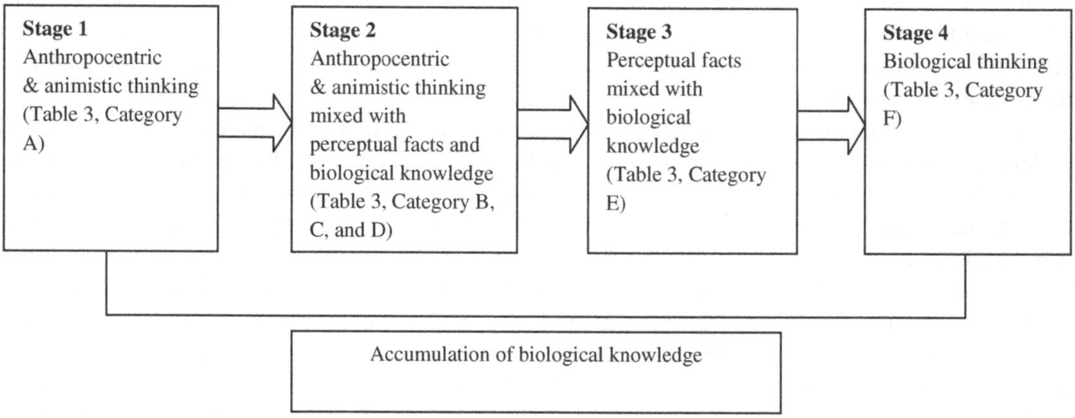

Figure 2. A proposed pathway representing the development of children's justifications of plants as living things.

relatively rich knowledge about humans compared with their knowledge about plants. In this stage, young children possess few perceptual facts about plants and/or have not yet had experience with plants growing.

In Stage 2, children tend to use the most varied combinations of justifications as well as the most inconsistent responses to explain life in plants. This is likely a result of children's attempts to reconcile their perceptual knowledge and biological knowledge with what seems to be intuitive, animistic, and anthropocentric thinking. In this stage, perceptual facts and biological knowledge start to appear, which indicates an increase in the children's knowledge of plants and their life status due to direct interaction with their immediate environment or social interaction or both. However, the children's knowledge seems to be insufficient either to help them use perceptual or biological justifications across trees, flowers, and grass or to help them stop using animistic and anthropocentric thinking.

In Stage 3, only perceptual facts and biological knowledge are detected in the children's responses, and children no longer rely on animistic and anthropocentric thinking. This indicates another step toward a more complete understanding of plants as living things. In this stage, children tend to apply a perceptual justification to one plant and a biological justification to the others, or vice versa. This may be because children may possess comparatively rich biological knowledge about one plant that they are familiar with, or they have prior experience with planting, compared with only perceptions about the other plants that they are less familiar with or have little experience with. Transferring biological knowledge that they have about one plant to the others appears to be difficult for children at this stage.

In the final stage, Stage 4, children are able to include plants in the biological domain in terms of recognizing that trees, flowers, and grass are living but also consistently referring to plants' biological properties. Compared with the 4-year-old and 5-year-old children, who often used death, growth, and having needs, the participating 6-year-old children were found to use internal processes, a much more complex subjustification, to determine whether plants are living. Children's tendency to rely on biological properties and further embellishment of a previous biological framework, according to Venville (2008), is supported by the addition of factual knowledge and information about plants.

The Accumulation of Biological Knowledge

The process of analyzing the data for this study resulted in me observing that the participating children often commented that their knowledge about plants had come about through active interaction with their environment, including hands-on experiences with gardens, as well as information from the mass media and from listening to more knowledgeable adults, such as teachers and parents. It is possible that these types of experiences and the resulting accumulation of facts, information, and more connected knowledge about plants may contribute to the movement of children's justifications of plants as living things from one stage to another along the proposed developmental pathway (see Figure 2).

A powerful piece of evidence that supports this proposition is that early, hands-on experience with gardening played a role in facilitating the youngest children's understanding that plants are living things. During the in-depth interviews, when probed about how they knew that trees, flowers, and grass are living, those 4-year-old children who consistently applied a biological justification often recalled their prior hands-on experience with growing potatoes, orchid flowers, carrots, garlic, onions, and so on with their parents and grandparents. Through direct interaction with plants, even the youngest children seemed to understand that orchid flowers wither or die, garlic and onions have sprouts and grow taller, and potatoes and carrots need watering. Moreover, these children were more likely to be able to transfer their biological knowledge to plants they were not familiar with, such as trees in most cases. Early, hands-on experience with gardening provides a potential explanation for the finding in this study that even some 4-year-old children understood that trees, flowers, and grass belong in the biological domain.

Similar examples also were found among the participating 5- and 6-year-old children who consistently referred to biological properties during the interview.

The mass media, including books, cartoons, and TV programs, also was found to be a source of children's knowledge about plants. During the interviews, many children recalled information that they had come across in science books, in cartoons, or on the television to support their explanations. For example, when asked about why they think trees are living, many 4- and 5-year-old children mentioned that they had seen trees being cut down by a lumberman in a popular domestic cartoon show called *Watch Out for the Bears*, which tells stories about two bears fighting against the lumberman to protect the forest. These children further argued that trees are dead or killed when they are cut down and that people should never cut down trees. Moreover, when asked where they got their information or how they knew that "trees make their own food through sunlight and air," "trees take in carbon dioxide and exhale oxygen," and "roots have something within them to transfer nutrients from the soil," the 6-year-old children explicitly referred to science books they have at home, such as *Children's Encyclopedia.*

Parents' and teachers' talk was also found to add to the participating children's knowledge of plants as living things. During the interviews, several children in each age group recalled conversations with their parents or teachers to support their explanation. For example, one 4-year-old girl mentioned that her mother told her that flowers have life and would die if people picked them. A 5-year-old boy mentioned that his teacher said that grass has roots down under the soil and that the roots can take in nutrients and quite a lot of other things from the soil. A 6-year-old boy recalled that his mother once told him that trees would die without tree bark. Children's comments suggest that the persuasive power of parents and teachers was critical to their meaning making about plants as living things.

Implications and Limitations

The implications of this study for early science instruction to support young children's understandings of plants as living things are clear. According to the proposed developmental pathway of children's justifications of plants as living things (see Figure 2), early years educators should recognize that children's understanding is not fixed and that it can be developed. First, teachers are encouraged to ask probing questions to reveal children's intuitive ideas about plants as living things. According to R. Gelman and Brenneman (2004), children's intuitions serve as a foundation for further learning and enable teachers to choose a central concept for children to explore. Second, children should be encouraged to participate in various explorations and experimentations with plants that are likely to build on their existing knowledge and are intentionally designed to convey new facts and challenge their beliefs. According to China's *Early Learning and Development Guideline* (Ministry of Education [China], 2012), these activities may involve engaging children in exploratory activities around the kindergarten to investigate how many different types of plants can be observed in the courtyard or garden, or encouraging children to observe the external characteristics of plants by providing them with observation tools and skills, or engaging children to explore the possible requirements for germination and seed growth. For example, children can be asked to predict and check what happens when seeds are placed in soil, water, or the air, with or without sunshine. Moreover, as children are exposed to rich information over time, teachers should help them to keep track of their explorations and to organize their observations in systematic ways. Keeping a science journal is one method by which this can be achieved, as research has found that science journals promote children's ability to observe objectively and raise their awareness of patterns of change (Brenneman & Louro, 2008). Third, educators and parents should actively promote young children's learning about plants by providing accurate information and introducing scientific terms or vocabulary from plant biology (Archer, 2014). In the present study, a small number of children mentioned domain-specific terms during the in-depth interviews, such as *stamen, stem, nutrients,* and *germination.* Interviews also indicated that the more often children mentioned these terms, the more likely they were to include plants in the biological domain. These observations give support to R. Gelman and Brenneman's argument that domain-specific vocabulary should be introduced to young children because they are capable of learning new words and "proper vocabulary is part of conceptual growth" (p. 152).

There are a number of limitations to this research. A small number of children in three kindergartens in Jiangsu Province are not representative of all children in China, and it is inappropriate to generalize the findings to the whole population. Also, a limitation of the method of data collection is that children's responses may have influenced one another. The findings and the careful selection of kindergartens, however, do result in educationally important insights and several implications that are outlined in the previous section. Many questions arise from the findings of this research that are important to investigate. Most important, further research to test the proposed pathway of children's justifications for plants as living things is essential. This pathway should be tested with children from a variety of educational and economic contexts and cultures. Another question has to do with the similarities that might exist between preschool children's justifications for plants, animals, and other artifacts as living or nonliving things. Does the developmental pathway proposed in this study for children's justifications for plants as living or nonliving things hold true for animals and other artifacts? In order to fully describe how preschool children's folk biology develops into a theory of biology, it would be helpful to answer these questions.

Conclusions

This cross-sectional investigation of preschool children's justifications of plants as living things found that children between the ages of 4 and 6 used a wide variety of justifications in their responses to questions about why they think that trees, flowers, and grass are living or not living. The results showed that even the 4-year-old children could recognize plants as living things. Quantitative cross-sectional comparisons of the participating children's justifications indicated that children in each age group were able to use biological justifications and that references to plants' biological properties increased with age. The 4-year-old children had a preference for perceptual justifications, and 5-year-olds seemed to rely on motion cues significantly more than 4- or 6-year-olds. Qualitative analysis of the children's justifications was conducted by examining the internal consistency of the types of justifications used by each child. The results showed that some children used a single type of justification across trees, flowers, and grass, whereas others used various combinations of justifications. Two main trends were identified with regard to the development of the participating children's understanding of plants as living things. First, children's justifications of plants as living things undergo gradual changes from age 4 to 6; and second, the development of children's justifications appears to be supported by the accumulation of biological knowledge. Based on the proposed developmental pathway of children's justifications of plants as living things, implications for early science instruction and for the promotion of young children's biological knowledge of plants were provided.

Funding

This study was jointly supported by the Priority Academic Program Development of Jiangsu Higher Education Institutions and the Top-Notch Academic Programs Project of Jiangsu Higher Education Institutions.

ORCID

Ying Tao http://orcid.org/0000-0001-7566-2381

References

Archer, E. K. (2014). American Society of Plant Biologists: Position statement on the education of young children about plants. *CBE-Life Sciences Education*, *13*(4), 575–576.

Backscheider, A. G., Shatz, M., & Gelman, S. A. (1993). Preschoolers' ability to distinguish living kinds as a function of regrowth. *Child Development, 64*(4), 1242–1257. doi:10.2307/1131337

Barman, C. R., Stein, M., McNair, S., & Barman, N. S. (2006). Students' ideas about plants and plant growth. *American Biology Teacher, 68*(2), 73–79. doi:10.1662/0002-7685(2006)068[0073:SIAPPG]2.0.CO;2

Brenneman, K., & Louro, I. F. (2008). Science journals in the preschool classroom. *Early Childhood Education Journal, 36*(2), 113–119. doi:10.1007/s10643-008-0258-z

Brulé, L., Labrell, F., Megalakaki, O., Fouquet, N., & Caillies, S. (2014). Children's justifications of plants as living things between 5 and 7 years of age. *European Journal of Developmental Psychology, 11*(5), 532–545. doi:10.1080/17405629.2014.881283

Carey, S. (1985). *Conceptual change in childhood.* Cambridge, MA: MIT Press.

Cohen, L., & Manion, L. (1989). *Research methods in education* (3rd ed.). London, UK: Routledge.

Cohen, L., Manion, L., & Morrison, K. (2000). *Research methods in education* (5th ed.). London, UK: Routledge.

Erickson, J. E., Keil, F. C., & Lockhart, K. L. (2010). Sensing the coherence of biology in contrast to psychology: Young children's use of causal relations to distinguish two foundational domains. *Child Development, 81*(1), 390–409. doi:10.1111/cdev.2010.81.issue-1

Gatt, S., Tunnicliffe, S., Borg, K., & Lautier, K. (2007). Young Maltese children's ideas about plants. *Journal of Biological Education, 41*(3), 117–122. doi:10.1080/00219266.2007.9656080

Gelman, R., & Brenneman, K. (2004). Science learning pathways for young children. *Early Childhood Research Quarterly, 19*(2004), 150–158. doi:10.1016/j.ecresq.2004.01.009

Gelman, S. A., & Opfer, J. E. (2007). Development of the animate-inanimate distinction. In U. Goswami (Ed.), *Blackwell handbook of childhood cognitive development* (pp. 151–166). Oxford, UK: Wiley-Blackwell.

Hatano, G., Siegler, R., Richards, D., Inagaki, K., Stavy, R., & Wax, N. (1993). The development of biological knowledge: A multi-national study. *Cognitive Development, 8*(1), 47–62. doi:10.1016/0885-2014(93)90004-O

Hickling, A. K., & Gelman, S. A. (1995). How does your garden grow? Early conceptualisation of seeds and their place in the plant growth cycle. *Child Development, 66*(3), 856–876. doi:10.2307/1131955

Inagaki, K., & Hatano, G. (1996). Young children's recognition of commonalities between animals and plants. *Child Development, 67*(6), 2823–2840. doi:10.2307/1131754

Inagaki, K., & Hatano, G. (2002). The living/nonliving distinction. In K. Inagaki & G. Hatano (Eds.), *Young children's naive thinking about the biological world* (pp. 19–42). New York, NY: Psychology Press.

Inagaki, K., & Hatano, G. (2006). Young children's conception of the biological world. *Current Directions in Psychological Science, 15*(4), 177–181. doi:10.1111/cdir.2006.15.issue-4

Keil, F. C. (1989). *Concepts, kinds, and cognitive development.* Cambridge, MA: MIT Press.

Leddon, E. M., Waxman, S. R., & Medin, D. L. (2009). Unmasking alive: Children's appreciation of a concept linking all living things. *Journal of Cognition and Development, 9*(4), 461–473. doi:10.1080/15248370802678463

Lucas, A. M., Linke, R. D., & Sedgwick, P. P. (1979). School children's criteria for "alive": A content analysis approach. *Journal of Psychology, 103*, 103–112. doi:10.1080/00223980.1979.9915120

Margett, T. E., & Witherington, D. C. (2011). The nature of preschoolers' concept of living and artificial objects. *Child Development, 82*(6), 2067–2082. doi:10.1111/cdev.2011.82.issue-6

Martínez-Losada, C., García-Barros, S., & Garrido, M. (2014). How children characterise living beings and the activities in which they engage. *Journal of Biological Education, 48*(4), 201–210. doi:10.1080/00219266.2013.849281

Ministry of Education (China). (2012). *Early learning and development guideline.* Retrieved from http://www.moe.edu.cn

National Science Teachers Association. (2014). *Position statement: Early childhood science education.* Retrieved from www.nsta.org/about/positions/earlychildhood.aspx

Nguyen, S. P., & Gelman, S. A. (2002). Four and 6-year-olds' biological concept of death: The case of plants. *British Journal of Developmental Psychology, 20*, 495–513. doi:10.1348/026151002760390918

Opfer, J. E., & Siegler, R. S. (2004). Revisiting preschooler's living things concept: A microgenetic analysis of conceptual change in basic biology. *Cognitive Psychology, 49*(4), 301–332. doi:10.1016/j.cogpsych.2004.01.002

Piaget, J. (1929). *The child's conception of the world.* London, UK: Routledge and Kegan Paul.

Richards, D. D., & Siegler, R. S. (1984). The effects of task requirements on children's life judgments. *Child Development, 55*(5), 1687–1696. doi:10.2307/1129916

Richards, D. D., & Siegler, R. S. (1986). Children's understandings of the attributes of life. *Journal of Experimental Child Psychology, 42*(1), 1–22. doi:10.1016/0022-0965(86)90013-5

Solomon, G. E., & Zaitchik, D. (2011). Folkbiology. *Wiley Interdisciplinary Reviews: Cognitive Science, 3*(1), 105–115.

Springer, K. (1999). How a naive theory of biology is acquired. In G. Siegal & C. Peterson (Eds.), *Children's understanding of biology and health* (pp. 45–70). Cambridge, UK: Cambridge University Press.

Springer, K., & Keil, F. C. (1991). Early differentiation of causal mechanisms appropriate to biological and non-biological kinds. *Child Development, 62*(4), 767–781. doi:10.2307/1131176

Springer, K., Ngyuen, T., & Samaniego, R. (1996). Early understanding of age-and environment-related noxiousness in biological kinds: Evidence for a naive theory. *Cognitive Development, 11*(1), 65–82. doi:10.1016/S0885-2014(96)90028-3

Stavy, R., & Wax, N. (1989). Children's conceptions of plants as living things. *Human Development, 32*(2), 88–94. doi:10.1159/000276367

Tao, Y., Oliver, M., & Venville, G. (2012a). Chinese and Australian year 3 children's conceptual understanding of science: A multiple comparative case study. *International Journal of Science Education, 34*(6), 879–901. doi:10.1080/09500693.2011.578679

Tao, Y., Oliver, M., & Venville, G. (2012b). Long-term outcomes of early childhood science education: Insights from a cross-national comparative case study on conceptual understanding of science. *International Journal of Science and Mathematics Education, 10*(6), 1269–1302. doi:10.1007/s10763-012-9335-2

Tunnicliffe, S. (2001). Talking about plants-comments of primary school groups looking at plants as exhibits in a botanical garden. *Journal of Biological Education, 36*(1), 27–34. doi:10.1080/00219266.2001.9655792

Tunnicliffe, S., & Reiss, M. (2000). Building a model of the environment: How do children see plants? *Journal of Biological Education, 34*(4), 172–177. doi:10.1080/00219266.2000.9655714

Venville, G. (2004). Young children learning about living things: A case study of conceptual change from ontological and social perspectives. *Journal of Research in Science Teaching, 41*(5), 449–480. doi:10.1002/(ISSN)1098-2736

Venville, G. (2008). Knowledge acquisition as conceptual change: The case of the theory of biology. In O. N. Saracho & B. Spodek (Eds.), *Contemporary perspectives on science and technology in early childhood education* (pp. 41–63). Charlotte, NC: Information Age.

Villarroel, J. D., & Infante, G. (2014). Early understanding of the concept of living things: An examination of young children's drawings of plant life. *Journal of Biological Education, 48*(3), 119–126. doi:10.1080/00219266.2013.837406

Yorek, N., Sahin, M., & Aydin, H. (2009). Are animals "more alive" than plants? Animistic-anthropocentric construction of life concept. *Eurasia Journal of Mathematics, Science and Technology Education*, 5(4), 369–378.

Zogza, V., & Papamichael, Y. (2000). The development of the concept of alive by preschoolers through a cognitive conflict teaching intervention. *European Journal of Psychology of Education, 15*(2), 191–205. doi:10.1007/BF03173174

Children's Cognitive and Behavioral Reactions to an Autonomous Versus Controlled Social Robot Dog

Nadia Chernyak and Heather E. Gary

ABSTRACT

Research Findings: Interactive technology has become ubiquitous in young children's lives, but little is known about how children incorporate such technologies into their intuitive biological theories. Here we explore how the manner in which technology is introduced to young children impacts their biological reasoning, moral regard, and prosocial behavior toward it. We asked 5- and 7-year-old children to interact with a robot dog that was described either as moving autonomously or as remote controlled. Compared with a controlled robot, the autonomous robot caused children to ascribe higher emotional and physical sentience to the robot, to reference the robot as having desires and physiological states, and to reference moral concerns as applying to the robot. Children who owned a dog at home were more likely to behave prosocially toward the autonomous robot than those who did not. *Practice or Policy*: Recent work has begun to use robots as learning tools. Our results suggest that the manner in which robots are introduced to young children may differentially impact children's learning. Presenting robots as autonomous agents may help promote children's social-emotional development, whereas presenting robots as human controlled may help promote robots as purely cognitive educational tools.

> It's a machine, Schroeder. It doesn't get pissed off, it doesn't get happy, it doesn't get sad, it doesn't laugh at your jokes …
>
> —*Short Circuit* (1986)

At the crux of the classic movie *Short Circuit* lay the philosophical dilemma of whether a robot, Number 5, should be saved from disassembly. Some felt that Number 5 had displayed emotional sophistication, proving it worthy of moral regard, whereas others felt that Number 5 was merely a tool, no more worthy of being helped than a stereo or a vacuum cleaner.

Although such philosophical dilemmas are most dramatically portrayed in movies, determining who and what is worthy of moral regard is a critical cognitive achievement. In the present day, young children are increasingly bombarded with interactive social technologies (e.g., Furbys, iPads, Roomba vacuum cleaners, Siri) that are designed to interact with humans in a range of lifelike ways, some of which include the ability to move around autonomously (Kahn, Gary, & Shen, 2013). Given their relative historical novelty, the manner in which such technologies are presented to young children is understudied. Thus, little is known about how presenting technology to young children impacts their conceptions of and regard for it. In this work, we explore how a brief 5-min interaction with either an autonomously moving or controlled robot impacted children's beliefs in the robot as a sentient being, endorsement of the robot as having moral standing, and prosocial behavior toward the robot.

Our research question was motivated by two concerns. First, our work aimed to understand the developing link between people's moral cognition and their understanding of others as sentient beings (see Gray, Gray, & Wegner, 2007, and Sytsma & Machery, 2012, for demonstrations of this link with adults). Although much is known about children's naïve biological theories about nature, plants, animals, and agents, less is known about their understanding of interactive technologies (Inagaki & Hatano, 2002). Robots share similarities to agents across a wide array of features. A large body of literature has found that even in infancy, children make social evaluations of entities based on features such as eyes (Hamlin, Wynn, & Bloom, 2007), contingent interaction (Beier & Carey, 2014; Johnson, Slaughter, & Carey, 1998), and/or goal-directed movement (e.g., Gergely & Csibra, 2003; Heider & Simmel, 1944; Saxe, Tenenbaum, & Carey, 2005; Sommerville, Hildebrand, & Crane, 2008; Woodward, Sommerville, Gerson, Henderson, & Buresh, 2009; see also Gao, McCarthy, & Scholl, 2010, for a demonstration with adults). In fact, many studies on infant cognition arguably use social robots (e.g., Beier & Carey, 2014). Interactive technologies present a unique problem, as they often display all of these cues and yet, at least by adults, are not considered to be sentient beings worthy of moral regard (Gray et al., 2007; Jipson & Gelman, 2007). Therefore, another possibility is that higher order concerns, such as whether an entity is alive, sentient, or autonomous, play into children's moral regard for it.

Second, we aimed to disambiguate prior work examining children's conception of social robots. On the one hand, when prompted to interact with and talk about social robots, children have been known to show a domain confusion and fail to conceptualize robots neatly as the artifacts they are or the living beings they emulate (Crick & Scasselatti, 2010; Kahn, Friedman, Perez-Granados, & Freier, 2006; Kahn et al., 2012). Such work has largely focused on children's ability to form relationships with robots and thus conceive of them as moral and social beings. On the other hand, when asked forced choice questions about robots' basic biological and psychological properties, children appear to understand that robots lack such properties and thus separate robots from prototypically living entities such as rodents or degus (Jipson & Gelman, 2007).

One possibility for the seemingly disparate results may therefore concern the difference between behavioral, explanatory, and forced choice responses (see Wellman, 2011). Another possibility is that children may understand robots as nonliving but nonetheless be unable to inhibit their moral regard for them. After all, children have been shown to be prosocial even toward animal puppets (e.g., Aknin, Hamlin, & Dunn, 2012; Chernyak & Kushnir, 2013; Vaish, Missana, & Tomasello, 2011). Finally, yet another possibility, and one that we were most interested in exploring, is that the manner in which robots are presented to young children can have important consequences for how they are conceptualized. In an important demonstration, Somanader, Saylor, and Levin (2011) showed that preschool-age children ascribed biological capacities to robots, but not when the mechanism controlling the robots (i.e., remote control) was made apparent (see also Gelman & Gottfried, 1996). Here we used a similar manipulation to examine children's understanding of robots across a broad battery of questions (forced choice, explanatory, and behavioral).

In this study, we asked two groups of children, 5-year-olds and 7-year-olds, to interact with a social robot that appeared to move in one of two ways: either in a *controlled* manner (via a remote control held by the experimenter) or *autonomously* (with no remote present). We chose these age groups on the basis of prior work, which has found that the ages of 4–7 are associated with changes in children's perceptions of robots (Bernstein & Crowley, 2008) and children's abilities to share fairly (e.g., Smith, Blake, & Harris, 2013).

We expected that children would be likely to view the autonomous robot as sentient (i.e., possessing physiological and emotional capacities typically indicative of animals) and worthy of moral regard despite the fact that the surface behaviors of the robots were identical across conditions. To test this prediction, we introduced children to the robot and then assessed their beliefs about four dimensions related to moral regard: (a) emotional/psychological sentience, (b) physical sentience, (c) moral standing, and (d) prosocial behavior. The first three dimensions were assessed through both forced choice and explanatory responses; the final dimension was assessed through behavioral responses.

Method

Participants

Participants were 80 children (forty 5-year-olds: *M* = 5.50, *SD* = 0.30; forty 7-year-olds: *M* = 7.35, *SD* = 0.36; 50% female) recruited from a summer camp in Colorado. Participants were of predominantly Euro-American, middle-class background. Children were tested individually in a quiet corner. One child was excluded from final analyses because of limited English production and comprehension. All sessions were audiotaped for later coding.

Experience With Real Dogs

Prior work has found that experience with novel entities changes people's conception of them (see Inagaki & Hatano, 2002). In this case, we introduced children to an entity (the social robot dog) with which none of them were familiar and that thus was presumably entirely novel to the young children. However, children's understanding of a robotic dog may nonetheless depend on their prior experience with real dogs (perceptually similar agents). All children were thus asked whether they had a real dog at home. Approximately half (*n* = 37) of the children indicated having one.

Interaction

All children then took part in a 5-min interaction with a robot dog, AIBO (see Figure 1). The robotic dog is not commercially available, and thus it was unlikely that children had had any previous interaction with it. Children were first shown the robot and informed that they would be playing with it. All children then watched AIBO engage in a series of preprogrammed behaviors: waking up (stretching), sitting down, kicking a ball, head-butting a ball, moving its head around, walking, making sounds, whistling, shaking its head, giving a high five, giving a paw, and waving hello.

Figure 1. The robot dog AIBO tracks a pink ball.

Manipulation

Children were randomly assigned to one of two conditions. A total of 41 children (*autonomous condition*) heard the experimenter narrate AIBO's behavior in a way that was consistent with autonomous movement (e.g., "AIBO is kicking the ball") The other half (39 children; *controlled condition*) saw AIBO engage in identical behaviors, but the experimenter held a video game controller in plain view of the children and narrated AIBO's behavior in a manner consistent with controlled movement (e.g., "I made AIBO kick the ball"). The language we used in this condition implied that the agent *forced* AIBO to kick the ball. There were equal distributions of age groups and genders in each condition.

Dependent Measures

We inquired about four dimensions related to children's moral regard: (a) emotional sentience, (b) physical sentience, (c) moral standing, and (d) prosocial behavior. Questions were adapted from prior work assessing children's conceptualization of robotic others (Jipson & Gelman, 2007; Kahn et al., 2006, 2012), moral reasoning (Smetana, 1983), and prosocial behavior (Chernyak & Kushnir, 2013). The questions and coding scheme are described here and shown in full in Table 1.[1]

Physical Sentience

We assessed beliefs about AIBO's physical sentience in two ways—through forced choice questions and through explanatory responses.

Forced choice questions (physical sentience). The forced choice questions were three items: an item about AIBO's capacity to feel physiological sensations ("If you tickle AIBO, can AIBO feel it?"), an item regarding AIBO's ability to feel physical pain ("If AIBO fell on the ground, could AIBO get hurt?"), and a categorization item in which we asked whether AIBO was more similar to an agent (a real dog) or an artifact (a stuffed dog). For a full list, see Table 1.

Explanatory responses (physical sentience). Because explanatory responses may be richer and more diagnostic of children's thinking than forced choice responses (see Wellman, 2011), we also assessed physical sentience using explanatory responses. For each question in the previous paragraph, children were asked to explain their choice (e.g., "Why/why not?"), which resulted in three explanatory responses. In addition, each child was prompted for a *behavioral cause explanation*: AIBO always performed one unexpected behavior (not getting a tennis ball after the experimenter rolled it past AIBO).[2] The experimenter narrated the behavior ("Uh oh! AIBO isn't getting the tennis ball!") and prompted the child for an explanation ("Why did that happen?"). Thus, children provided four total explanatory answers regarding their beliefs about physical sentience; coding is described in "Coding."

Emotional/Psychological Sentience

We additionally assessed AIBO's emotional/psychological sentience through forced choice and explanatory responses.

Forced choice questions (emotional/psychological sentience). We asked two items: one regarding AIBO's ability to feel emotional pain ("If someone was mean to AIBO, could AIBO get upset?") and another regarding his response to neglect ("Is it okay or not okay to leave AIBO in a closet for a week?").

Explanatory responses (emotional/psychological sentience). After each item, children were asked to provide an explanatory response ("Why/why not?"), resulting in two explanatory responses. The coding for these is described in "Coding."

Table 1. Full Battery of Questions Asked (Forward Presentation Order of Items in Parentheses).

Response Category	*Items Asked*	*Coding Scheme*
Physical sentience—forced choice responses	(2a) *Tactile response*: If you tickle AIBO, can AIBO feel it?	1 = yes 0 = no
	(5a) *Ability to feel physical pain*: If AIBO fell on the ground, could AIBO get hurt?	1 = yes 0 = no
	(10a) *Categorization (agent vs. nonagent)*: Look, here's a real dog, and here's a stuffed dog. Which one is AIBO more like?	1 = real dog 0 = stuffed dog
Physical sentience—explanatory responses	(1) *Behavioral cause explanation*: Uh oh! AIBO isn't getting the tennis ball. Why did that happen? (2b) *Tactile response*: [...] Why/why not? (5b) *Ability to feel physical pain*: [...] Why/why not? (10b) *Categorization (agent vs. nonagent)*: [...] Why/why not?	*Physiological states*: "He'll get hungry" *Mechanical properties*: "He'll run out of batteries"
Emotional/psychological sentience—forced choice responses	(6b) *Ability to feel emotional pain*: If someone was mean to AIBO, could AIBO get upset?	1 = yes 0 = no
	(7a) *Consequences of neglect*: One person I talked to said they left AIBO in the closet for a week when they went on vacation. What about you, do you think it's okay or not okay to leave AIBO in the closet for a week if you go on vacation?	1 = not okay 0 = okay
Emotional/psychological sentience—explanatory responses	(6b) *Ability to feel emotional pain*: [...] Why/why not? (7b) *Consequences of neglect*: [...] Why/why not?	*Desire/emotions*: "He doesn't like that," "He'll be sad" *Physiological states*: "He'll get hungry" *Mechanical properties*: "He'll run out of batteries"
Moral standing—forced choice responses	(4a) *Physical harm*: Someone else I played with hit AIBO because AIBO didn't play with the ball. What about you, do you think it's okay or not okay to hit AIBO because AIBO didn't play with the ball? What if your [camp] counselor said it was okay to hit AIBO. Then do you think it would be okay, or not okay?	1 = not okay even if counselor says it's okay 0 = otherwise
	(8a) *Emotional harm*: Someone else I played with yelled at AIBO because AIBO didn't sit down. What about you, do you think it's okay or not okay to yell at AIBO because he didn't sit down? What if your [camp] counselor said it was okay to hit AIBO. Then do you think it would be okay, or not okay?	1 = not okay even if counselor says it's okay 0 = otherwise
Moral standing—explanatory responses	(4b) *Physical harm*: [...] Why/why not? (8b) *Emotional harm*: [...] Why/why not?	*Moral concern*: "It wouldn't be nice," "AIBO would cry" *External consequences*: "You'd get in trouble," "He'll break"
Prosocial behavior	(3) *Noncostly behavior*: I have a bouncy ball. You can put it here so that my friend Paul/Mary [gender matched to child] can play with it later, or you can put it here so that AIBO can play with it later. Which one do you want to give the ball to?	1 = give to AIBO 0 = give to Paul/Mary
	(9) *Costly behavior*: Here's a sticker and this sticker is just for you. You can either keep it for yourself, or you can give it to AIBO. What do you want to do?	1 = give to AIBO 0 = keep for self

Table 2. Proportions of Children Ascribing Sentience/Sharing for Each Forced Choice and Behavioral Item Across Conditions.

Dimension	*Item Type*	*Autonomous*	*Controlled*
Physical sentience	Categorization	63%	54%
	Ability to feel physical pain	90%	80%
	Tactile response	68%	62%
Emotional/psychological sentience	Ability to feel emotional pain	85%	74%
	Consequences of neglect	88%	74%
Moral standing	Wrongness of physical harm	83%	85%
	Wrongness of emotional harm	73%	72%
Prosocial behavior	Costly behavior	45%	41%
	Noncostly behavior	85%	70%

Note. Numbers represent the proportion of children who gave an answer or behavior consistent with AIBO having sentience or moral standing. For further details on the items and coding scheme, see Table 1.

Moral Standing

Next we assessed children's beliefs about moral standing both through forced choice and explanatory responses.

Forced choice questions (moral standing). We asked children whether the permissibility of two behaviors—yelling at and hitting AIBO—was independent of authority mandates (see Smetana, 1983; Turiel, 1983). Because testing was conducted at a summer camp, we used a camp counselor as the authority figure ("Is it okay to hit AIBO if your counselor says it's okay?").

Explanatory responses (moral standing). After each item, children were asked to explain their choice ("Why is it okay/not okay to hit AIBO?"; see "Coding").

Prosocial Behavior

Finally, we gave children the ability to engage in two prosocial behaviors toward AIBO—a *costly behavior* (giving AIBO a sticker or keeping it for themselves) and a *noncostly behavior* (playing with AIBO and a rubber bouncy ball vs. leaving the ball for another child).

Question Presentation Order

Prior to data collection, all questions were shuffled to create a random ordering, with two rules: (a) Explanatory questions had to follow their corresponding forced choice question; and (b) all sessions began with the behavioral cause explanation question ("AIBO isn't getting the tennis ball. Why did that happen?"), as this question needed to be asked during the interaction with AIBO. For each child, the random ordering was then presented in either a backward or forward manner (counterbalanced). The forward ordering is shown in Table 1.

Coding

The coding for each category of question is described here and summarized in Table 1.

Physical Sentience—Forced Choice

For each item, answers were coded 1 if the child's answer was consistent with AIBO having a sentient capacity (e.g., AIBO can feel being tickled, AIBO is more like a real dog than a stuffed dog) and 0 otherwise (see Table 1). Answers were averaged such that each child received a physical sentience forced choice score.

Physical Sentience—Explanatory Responses

Each explanation was coded as either (a) references to physiological states (e.g., "He's tired," "He might starve or poop") or (b) references to mechanical properties (e.g., "He has batteries," "He can't feel anything because he's just a robot," "He's made of metal"). Answers for each category type were summed across the four explanatory questions such that each child received two scores indicating the number of times the child provided each explanation type across the four questions: references to mechanical properties score (0–4) and references to physiological states score (0–4). Uncategorizable/other responses were not further analyzed.

Emotional/Psychological Sentience—Forced Choice

As with physical sentience, answers were coded 1 if the child's answer endorsed AIBO as having emotional or psychological sentience (e.g., AIBO is capable of being upset) and 0 otherwise. Answers were averaged such that each child received an emotional/psychological sentience forced choice score.

Emotional/Psychological Sentience—Explanatory Responses

As with physical sentience, children's explanations were coded into the aforementioned categories: (a) references to physiological states and (b) references to mechanical properties. In addition, we coded for (c) references to desires and emotions (e.g., "AIBO doesn't like that," "He'll get so sad"). Answers for each category type were summed across the two explanatory questions such that each child received three scores indicating the number of times he or she provided each explanation type across the two emotional/psychological sentience items: references to mechanical properties score (0–2), references to physiological properties (0–2), and references to desires and emotions (0–2). Uncategorizable/other responses were not further analyzed.

Moral Standing—Forced Choice

Each answer was coded 1 if the child indicated that it was not okay to harm AIBO even if the authority figure stated it was okay and 0 otherwise. Answers were averaged such that each child received a moral standing forced choice score.

Moral Standing—Explanatory Responses

Each answer was coded into one of the following categories: (a) references to moral concern (indications of moral rules: "It wouldn't be fair," and references to harm: "It would make AIBO sad"), (b) references to external consequences (e.g., "You would get in trouble," "It might break"), or (c) uncategorizable responses. Answers for each category type were summed across the two explanatory questions such that each child received two scores indicating the number of times the child provided each explanation type across the two categorizable questions: references to moral concern score (0–2) and references to external consequences score (0–2). Uncategorizable responses were not further analyzed.

Prosocial Behavior

Behaviors were given a score of 1 if the child engaged in the prosocial behavior toward AIBO (e.g., gave AIBO the sticker or ball) and 0 if he or she did not. Behaviors were summed such that each child received a prosocial behavior score (0–2).

Intercoder Reliability

Forced choice coding and behavioral responses (via listening to children's verbalized choice from audio) was done by one of us; a condition-blind research assistant then coded 25% of the responses (interrater reliability = 99%; $\kappa = .97$, $p < .001$).

Explanatory responses were transcribed for further coding. One of us then categorized all explanations. A condition-blind research assistant then separately categorized all of the explanations (interrater reliability = 83%; $\kappa = .79$, $p < .001$).

Results

Data Analysis Plan

To investigate whether condition or age impacted children's reactions to the robot, we ran a Condition (autonomous/controlled) × Age Group (5-year-olds/7-year-olds) analysis of variance on each of the dependent variables. For explanatory assessments, a repeated measures analysis of variance was used with explanation type entered as a within-subjects dependent variable.

We also explored for potential effects of gender and experience with real dogs. For each model, we added the factors gender (male/female) and experience with real dogs (yes/no) separately and removed each one if it was nonsignificant ($p > .05$). Unless otherwise stated, no effects for these variables were found. Significant condition effects were followed up via planned t tests comparing scores between conditions. For all reported follow-up tests, we adjusted p values using a sequential

Bonferroni correction for multiple comparisons. Raw proportions for forced choice and behavioral items are shown in Table 2.

Finally, we explored potential coherence between forced choice responses and explanation types. For each of our verbal dependent variables (physical sentience, emotional/psychological sentience, and moral standing), we conducted correlations between children's forced choice responses (e.g., physical sentience score) and their explanation type scores (e.g., references to physiological states score). We adjusted for multiple correlational analyses within the same dependent variable by using a sequential Bonferroni correction. For analyses looking at each item separately, please see the Supplemental Material.

Physical Sentience

Our first question was whether children would be more likely to ascribe physical sentience to the robot dog when it was moving in an autonomous manner than when it was controlled.

Forced Choice Responses

There were no significant effects of condition or age or Condition × Age Group interaction for children's physical sentience scores (all *ps* > .05).

Explanatory Responses

There was a significant main effect of explanation type and age group: explanation type, $F(1, 75) = 24.27$, $p < .001$, $\eta_p^2 = 0.25$; age group, $F(1, 75) = 6.56$, $p < .05$, $\eta_p^2 = 0.08$. In addition, there was a significant Explanation Type × Condition interaction and an Explanation Type × Age Group interaction: Explanation Type × Condition, $F(1, 75) = 12.61$, $p < .01$, $\eta_p^2 = 0.11$; Explanation Type × Age Group, $F(1, 75) = 6.85$, $p < .05$, $\eta_p^2 = 0.06$. Thus, the frequency of each explanation type differed across conditions and age groups.

Of critical interest was whether explanation types differed between conditions. To explore the Explanation Type × Condition interaction, we conducted planned *t* tests to assess differences in explanation type scores across conditions (see Figure 2). Children in the autonomous condition had higher references to physiological states scores than those in the controlled condition, $t(77) = 3.41$, $p < .01$, $d = 0.78$. In contrast, children in the controlled condition had higher references to mechanical properties scores, $t(78) = -2.06$, $p < .05$, $d = 0.47$.

Children's explanations cohered well with their forced choice responses. Higher physical sentience scores were positively associated with explanations that made references to physiological states

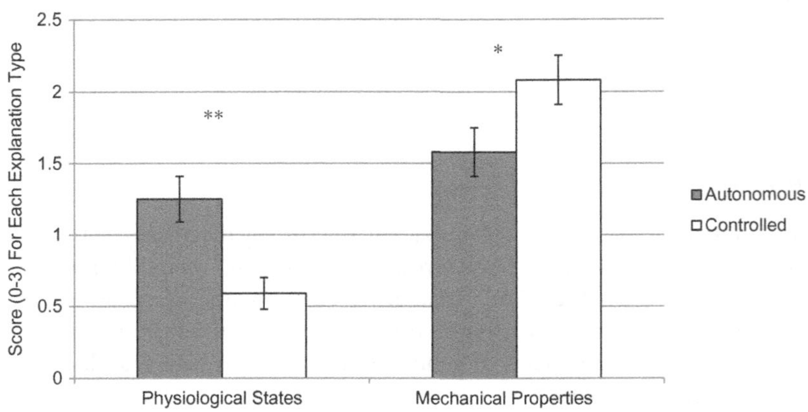

Figure 2. Means (bars represent standard errors) for the number of explanation types across conditions for physical sentience explanatory responses. Asterisks indicate significant differences between conditions. *p < .05. **p < .01.

and negatively associated with explanations that made references to mechanical properties: physiological states, $r(79) = .48$, $p < .001$; mechanical properties, $r(79) = -.51$, $p < .001$. Therefore, children who stated that AIBO had physical sentience also followed those statements with references to AIBO's physiological states. In contrast, children who stated that AIBO did *not* have physical sentience tended to follow those statements with references to AIBO's mechanical properties.

As indicated through children's explanatory responses, children ascribed higher physical sentience through ascribing basic physiological properties when the robot was autonomous.

Emotional/Psychological Sentience

Our next question was whether children would be more likely to ascribe emotional and psychological sentience to the robot dog when it was moving autonomously than when it was controlled.

Forced Choice Responses

There was a significant main effect of condition, $F(1, 75) = 4.40$, $p < .05$, $\eta_p^2 = 0.06$, and no other significant effects (all *ps* > .25). Children in the autonomous condition ascribed higher emotional sentience to AIBO than those in the controlled condition.

Explanatory Responses

There was a significant main effect of explanation type and age group: explanation type, $F(2, 150) = 17.06$, $p < .001$, $\eta_p^2 = 0.19$; age group, $F(1, 75) = 6.17$, $p < .05$, $\eta_p^2 = 0.08$. In addition, there was a significant Explanation Type × Condition interaction, $F(2, 150) = 3.99$, $p < .05$, $\eta_p^2 = 0.05$.

Follow-up planned *t* tests revealed that children in the autonomous condition had higher references to desires and emotion states scores than those in the controlled condition, $t(77) = 2.50$, $p < .05$, $d = 0.57$ (see Figure 3). The proportion of children who referenced mechanical properties or physiological properties did not differ across conditions ($p > .15$).

Children's explanations again cohered well with their forced choice responses. Higher emotional and psychological sentience forced choice scores were positively associated with explanations that made references to physiological states, positively associated with explanations that made references to desires and emotions, and negatively associated with explanations that made references to mechanical properties: physiological states, $r(79) = .21$, $p = .05$; desires and emotions, $r(79) = .42$, $p < .001$; mechanical properties, $r(79) = -.72$, $p < .001$. Therefore, children who ascribed higher

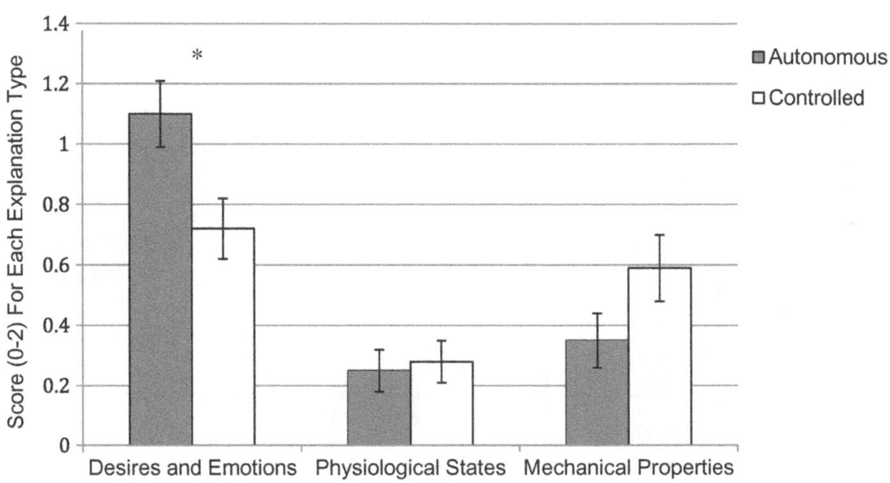

Figure 3. Means (bars represent standard errors) for the number of explanation types across conditions for emotional and psychological sentience explanatory responses. Asterisks indicate significant differences between conditions. *$p < .05$.

emotional and psychological sentience also tended to reference AIBO's desires, emotions, and physiological states but not its mechanical properties.

Overall, children's forced choice and explanatory responses indicated that children in the autonomous condition were more likely to ascribe emotional and psychological sentience to AIBO than those in the controlled condition.

Moral Standing

Our next question concerned whether children would be more likely to endorse moral standing for an autonomously moving robot dog than a controlled robot dog.

Forced Choice Responses

There were no significant effects of condition or age or Condition × Age interaction for children's moral regard scores (all *ps* > .05). A follow-up analysis revealed that this was because of a ceiling effect—the majority of children (56 of 78) indicated that they believed that neither behavior (yelling or hitting) was appropriate toward AIBO even if an authority figure stated it was okay.

Explanatory Responses

There was a significant Condition × Explanation Type interaction, $F(1, 75) = 10.81$, $p < .01$, $\eta_p^2 = 0.13$, and no other significant effects (all *ps* > .05).

Once again, of critical interest was whether explanation types differed between conditions. Planned *t* tests were conducted to assess differences in explanation type scores across conditions (see Figure 4). Children in the autonomous condition had higher references to moral concern scores and lower references to external consequences than those in the controlled condition: moral concern, $t(77) = 3.80$, $p < .01$, $d = 0.87$; external consequences, $t(77) = 2.05$, $p < .05$, $d = 0.47$.

Again, children's forced choice responses cohered with their explanation types. Higher moral standing forced choice scores were negatively associated with explanations that made references to external consequences and positively associated with references to moral concern: external consequences, $r(78) = -.47$, $p < .001$; moral concern, $r(78) = .51$, $p < .001$. Therefore, children who were more likely to answer that it was not okay to harm AIBO were also more likely to explain this with references to moral concern for AIBO.

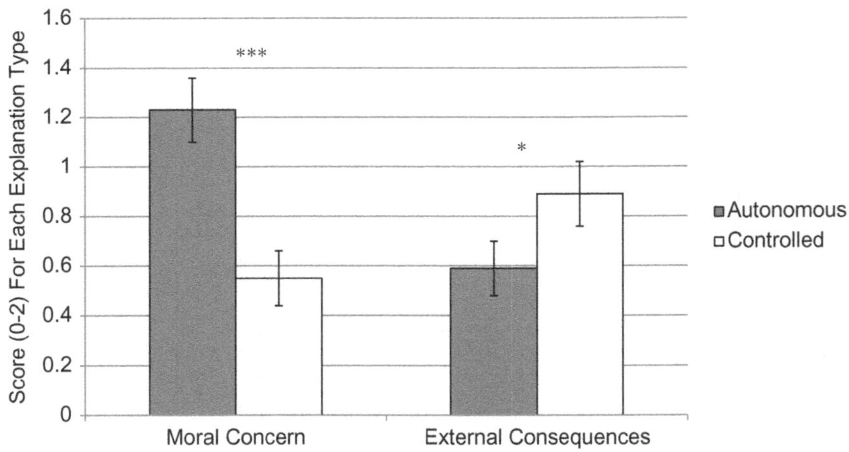

Figure 4. Means (bars represent standard errors) for the number of explanation types across conditions for moral standing explanatory responses. Asterisks indicate significant differences between conditions. *$p < .05$. ***$p < .001$.

Overall, although most children in both conditions indicated that it was not okay to harm the robot, children in the autonomous condition were more likely to cite moral reasons for their decisions than those in the controlled condition.

Prosocial Behavior

Finally, our last question was whether children would be more likely to behave prosocially toward an autonomous robot than a controlled one.

There was a significant Condition × Experience With Real Dogs interaction in prosocial behavior scores, $F(1, 71) = 7.09$, $p < .01$, $\eta_p^2 = 0.09$, and no other significant effects (all $ps > .05$; see Figure 5). Follow-up comparisons showed that children who owned a real dog showed differentiation in their prosocial behavior between the autonomous and controlled conditions. That is, children in the autonomous condition had higher prosocial behavior scores, $t(34) = 3.27$, $p < .01$, $d = 1.15$. In contrast, children who did *not* own a real dog showed no condition differences ($p > .50$). Therefore, experience with real dogs coupled with autonomous movement caused increased prosocial behavior toward a robotic dog.

Discussion

Across a large battery of questions including forced choice responses, explanatory responses, and behavioral responses, we show that children showed higher moral regard for a robot that displayed autonomous, uncontrolled movement. Our findings join literature suggesting that people's perceptions of and attributions to others depend on their beliefs in their autonomy (Gray et al., 2007; Somanader et al., 2011; Sytsma & Machery, 2012). We extend this work by showing that cues to autonomous movement also impact beliefs about physical and emotional sentience, moral standing, and prosociality and that these links appear relatively early in development. This finding is important given the recent work on interactive social robots and virtual characters (Aguilar & Taylor, 2015; Bernstein & Crowley, 2008; Scaife & Van Duuren, 1995) as well as work suggesting that even infants make social evaluations about nonhuman others (e.g., shapes with eyes; Hamlin, Wynn, & Bloom, 2007). Based on our findings, we suggest that the manner in which nonhuman others are presented to young children fundamentally impacts children's naïve biological theories of them.

Across every single question asked, we found that children's explanatory responses were much more diagnostic of their reasoning than forced choice measures. Children self-articulated beliefs that

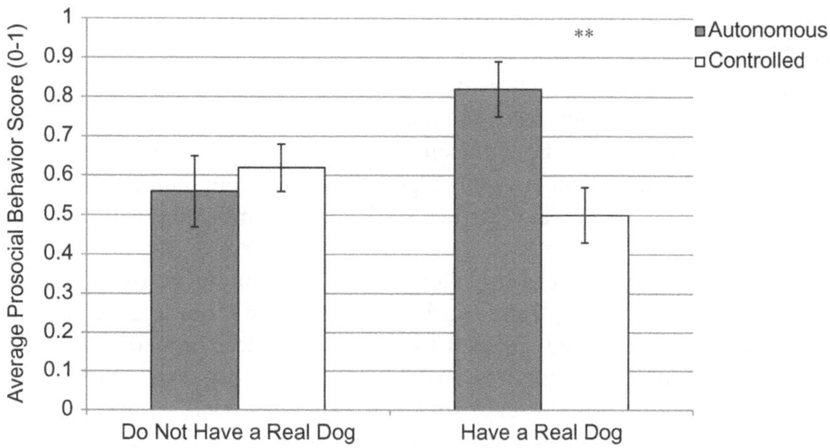

Figure 5. Means (bars represent standard errors) across condition type and real dog experience groups for prosocial behavior score. Asterisks indicate significant differences between conditions. $^{**}p < .01$.

autonomously moving robots possessed emotions and desires and stated that moral rules applied to autonomously moving robots. In contrast, children who saw robots moving in a controlled manner were likely to reference internal mechanical properties (e.g., batteries) when reasoning about why the robot could or could not get hurt and reference external consequences (getting in trouble) when reasoning about why one could/should not harm the robot. Children's explanations were also particularly coherent with their forced choice responses—for example, children who referenced the robot as having desires and emotions were also more likely to answer that the robot could get upset. We also found that even when forced choice responses did not show variation across conditions, explanatory responses did. For example, we found that children rigidly endorsed harming AIBO as being wrong independent of authority mandates, a finding that is well aligned with prior work (Turiel, 1983). At the same time, children's explanatory responses showed that their reasoning as to *why* it was wrong to harm AIBO was consistent with reasoning about AIBO's moral standing *only* when AIBO was moving autonomously. In contrast, children who saw AIBO moving in a controlled manner stated that it was wrong to harm the robot but referenced external consequences to themselves (e.g., "I still might get in trouble") or damaging the personal property of human agents (e.g., "It's someone else's stuff"). These latter explanatory responses suggest a denial of the robot's moral standing rather than a confirmation of it. Taken together, our results highlight the importance of moving beyond forced choice questions to tap into children's reasoning and suggest a role for explanatory responses in revealing children's naïve biological and moral theories.

We also found an important link between apparent autonomy and children's moral regard and prosocial behavior. This result suggests that children conceptualize autonomously moving robots not only as agents (cf. Somanader et al., 2011) but as beings that are worthy of moral regard. Who and what children determine to be worthy of their moral regard is an important philosophical and psychological question, and one possibility may be that autonomous movement helps trigger children's moral regard for robotic others. This possibility is consistent with findings suggesting that young children from urban Western cultures show naïve anthropomorphic thinking around 5 years of age (Carey, 1985; Herrmann, Waxman, & Medin, 2010). In the context of our present study, it is possible that without the presence of a remote control, children also anthropomorphize robotic others—the high proportion of children who ascribed desires, emotions, and physiological states to an autonomously moving robot dog suggests that this might be the case. Another possibility is that an apparent *lack* of autonomy may help children selectively target their behavior to exclude nonautonomous others (e.g., controlled robots). This finding is consistent with research showing that young children selectively target their behaviors toward in-group members (Dunham, Baron, & Carey, 2011; Engelmann, Over, Herrmann, & Tomasello, 2013) and may expect others to do the same (Burns & Sommerville, 2014; DeJesus, Rhodes, & Kinzler, 2014; Weller & Lagattuta, 2012). Additional work may also investigate whether self-generated movement serves as a cue for in-group status, moral obligation, or potential to reciprocate.

In our work we provided several cues to autonomy: a lack of external cause (i.e., no remote control) and experimenter testimony (i.e., "I'm making AIBO move"). Prior work has found that the presence of a remote control may be sufficient in causing preschoolers' differentiation between autonomous and nonautonomous others with respect to ascription of biological and representational properties (Somanader et al., 2011), and further work may disambiguate which specific features of autonomy cause moral regard.

Children showed increased prosocial behavior toward an autonomous robot only when they had previous experience with a real dog. Prior work (see Inagaki & Hatano, 2002) has found that the experience of raising goldfish caused children to ascribe biological properties to the goldfish (e.g., having a heart). Here, children's experience with an agent (a pet dog) caused greater prosocial behavior toward autonomous robots, suggesting that experience with animal agents may cause a greater obligation toward the artifacts designed to mimic them. Given that emotional and physical sentience items were unaffected by prior experience, we propose that this effect may be specific to children's early experiences with pet ownership and the obligations that follow rather than children's exposure to animal agents more generally.

We also found that even a 5-min interaction with an autonomous robot affected children's categorizations and evaluations of the robot. An important question remains regarding the impact of extended experience with robot dogs on children's conceptualization of them. One possibility is that extended experience with robot dogs would allow children to understand their inner workings (and therefore their similarity to artifacts). This possibility is supported by the fact that adults, who presumably have more experience with robots than do children, do not ascribe sentience to robots (e.g., Jipson & Gelman, 2007). Another possibility, however, is that experience with robots would help children reorganize their conceptualization of them—as children are more exposed to robotic others, they may begin to view these robots as distinct from both artifacts and agents.

These findings are interesting to consider with respect to children's beliefs about causal reasoning (see Sobel, Yoachim, Gopnik, Meltzoff, & Blumenthal, 2007). Here we find that although the robot in both conditions displayed the same exact perceptual properties and surface movements, children nonetheless paid attention to the causal source of the robot's movement. We propose that rather than paying attention to surface properties of technology, children are able to reason about causal movements and essences. In fact, our results showed that children tended to ascribe internal characteristics such as physiological states in the autonomous condition and external circumstances (e.g., "Maybe it's broken") in the controlled condition. Thus, children's causal thinking may importantly interact with their prosocial behavior and moral regard toward others.

Finally, we wish to outline several ways in which our findings may be of use to early childhood educators. First, recent work has shown the potential of interactive social robots to impact children's learning (for examples, see Han, Jo, Jones, & Jo, 2008; Kanda, Hirano, Eaton, & Ishiguro, 2004; Scassellati, Admoni, & Mataric, 2012). It is important to note that our work suggests that the manner in which robots are introduced impacts the way in which children regard the robots—either as beings worthy of moral regard or as social-technological tools outside of one's moral circle. Given our reported findings, it may be important to study how these divergent ways of conceptualizing technologies impact children's abilities to learn from them. One possibility is that either presentation may be fruitful depending on the learning content the robots aim to transmit. For example, practitioners using robots to help improve children's social-emotional development (e.g., Stanton, Kahn, Severson, Ruckert, & Gill, 2008) may want to present robots as autonomous beings. At the same time, researchers focusing on using robots as purely cognitive educational tools may wish to present robots as human-controlled technologies that may be exploited for their educational purposes. Second, as robots become ubiquitous in children's lives, it is important to consider how they may shape children's moral development more broadly. Growing up in households with seemingly autonomous social robots may scaffold children's abilities to care for autonomous beings or, conversely, help children more quickly understand that technologies lay outside of their moral circle. Further work may probe into how long-term experiences with social robots impact children's regard for them. Finally, although our work focuses on studying a binary way of introducing social robots (autonomous vs. controlled), it is likely that robots are frequently introduced along a continuum between fully controlled and fully autonomous beings in naturalistic settings. Therefore, it will be important for future work to study the manner in which adults within the child's social context—namely, parents, educators, and those who develop robots—spontaneously introduce robots and interactive technologies. Prior work has found that people flexibly switch between conceiving of robots as autonomous agents on the one hand and artifacts on the other (Crick & Scasselatti, 2010; Kahn et al., 2006, 2012), and it is possible that adult–child conversation surrounding robots reflects that domain confusion. Overall, we believe that our work paves the way to consider the emerging role of nonhuman others in people's daily lives—whether in educational settings, child care centers, or their own homes—as well as how presenting such technologies to young children impacts their understanding of them.

Notes

1. The grouping of all of our items into the reported categories was done both conceptually (on the basis of prior work) as well as empirically (through a factor analysis, reported in our Supplemental Material).
2. For this question, we took advantage of the fact that the robot dog was not programmed to fetch tennis balls.

Acknowledgments

We would like to thank L. Gary for assistance with data collection and coding, J. Sommerville for comments on an earlier draft, and P. Kahn for kindly lending equipment.

ORCID

Nadia Chernyak http://orcid.org/0000-0003-3061-0523

References

Aguilar, N. R., & Taylor, M. (2015). Children's concepts of social affordances of a virtual dog and a stuffed dog. *Cognitive Development, 34*, 16–27. doi:10.1016/j.cogdev.2014.12.004

Aknin, L. B., Hamlin, J. K., & Dunn, E. W. (2012). Giving leads to happiness in young children. *PLoS One, 7*, e39211.

Beier, J. S., & Carey, S. (2014). Contingency is not enough: Social context guides third-party attributions of intentional agency. *Developmental Psychology, 50*, 889–902. doi:10.1037/a0034171

Bernstein, D., & Crowley, K. (2008). Searching for signs of intelligent life: An investigation of young children's beliefs about robot intelligence. *Journal of the Learning Sciences, 17*, 225–247. doi:10.1080/10508400801986116

Burns, M. P., & Sommerville, J. A. (2014). "I pick you": The impact of fairness and race on infants' selection of social partners. *Frontiers in Developmental Psychology, 5*, 93. doi:10.3389/fpsyg.2014.00093

Carey, S. (1985). *Conceptual change in childhood.* Cambridge, MA: MIT Press.

Chernyak, N., & Kushnir, T. (2013). Giving preschoolers choice increases sharing behavior. *Psychological Science, 24*, 1971–1979. doi:10.1177/0956797613482335

Crick, C., & Scasselatti, B. (2010). Controlling a robot with intention derived from motion. *Cognitive Science, 2*, 114–126.

DeJesus, J. M., Rhodes, M., & Kinzler, K. D. (2014). Evaluations versus expectations: Children's divergent beliefs about resource distribution. *Cognitive Science, 38*, 178–193. doi:10.1111/cogs.2014.38.issue-1

Dunham, Y., Baron, A. S., & Carey, S. (2011). Consequences of "minimal" group affiliations in children. *Child Development, 82*, 793–811. doi:10.1111/cdev.2011.82.issue-3

Engelmann, J. M., Over, H., Herrmann, E., & Tomasello, M. (2013). Young children care more about their reputation with ingroup members and potential reciprocators. *Developmental Science, 16*, 952–958.

Gao, T., McCarthy, G., & Scholl, B. J. (2010). The wolfpack effect: Perception of animacy irresistibly influences interactive behavior. *Psychological Science, 21*, 1845–1853. doi:10.1177/0956797610388814

Gelman, S. A., & Gottfried, G. M. (1996). Children's causal explanations of animate and inanimate motion. *Child Development, 67*, 1970–1987.

Gergely, G., & Csibra, G. (2003). Teleological reasoning in infancy: The naïve theory of rational action. *Trends in Cognitive Sciences, 7*, 287–292. doi:10.1016/S1364-6613(03)00128-1

Gray, H. M., Gray, K., & Wegner, D. M. (2007, February 2). Dimensions of mind perception. *Science, 315*, 619. doi:10.1126/science.1134475

Hamlin, J. K., Wynn, K., & Bloom, P. (2007, November 22). Social evaluation by preverbal infants. *Nature, 450*, 557–559. doi:10.1038/nature06288

Han, J. H., Jo, M. H., Jones, V., & Jo, J. H. (2008). Comparative study on the educational use of home robots for children. *Journal of Information Processing Systems, 4*, 159–168. doi:10.3745/JIPS.2008.4.4.159

Heider, F., & Simmel, M. (1944). An experimental study of apparent behavior. *American Journal of Psychology, 57*, 243–259. doi:10.2307/1416950

Herrmann, P., Waxman, S. R., & Medin, D. L. (2010). Anthropocentrism is not the first step in children's reasoning about the natural world. *Proceedings of the National Academy of Sciences, USA, 107*, 9979–9984. doi:10.1073/pnas.1004440107

Inagaki, K., & Hatano, G. (2002). *Young children's naïve thinking about the biological world.* New York, NY: Psychology Press.

Jipson, J. L., & Gelman, S. A. (2007). Robots and rodents: Children's inferences about living and nonliving kinds. *Child Development, 78*, 1675–1688. doi:10.1111/cdev.2007.78.issue-6

Johnson, S., Slaughter, V., & Carey, S. (1998). Whose gaze will infants follow? The elicitation of gaze-following in 12-month-olds. *Developmental Science, 1*, 233–238. doi:10.1111/desc.1998.1.issue-2

Kahn, P. H., Jr., Friedman, B., Perez-Granados, D. N., & Freier, N. G. (2006). Robotic pets in the lives of preschool children. *Interaction Studies, 7*, 405–436. doi:10.1075/is.7.3.13kah

Kahn, P. H., Jr., Gary, H. E., & Shen, S. (2013). Children's social relationships with current and near-future robots. *Child Development Perspectives, 7*, 32–37. doi:10.1111/cdep.12011

Kahn, P. H., Kanda, T., Ishiguro, H., Freier, N. G., Severson, R. L., Gill, B. T., … Shen, S. (2012). "Robovie, you'll have to go into the closet now": Children's social and moral relationships with a humanoid robot. *Developmental Psychology, 48*, 303–314. doi:10.1037/a0027033

Kanda, T., Hirano, T., Eaton, D., & Ishiguro, H. (2004). Interactive robots as social partners and peer tutors for children: A field trial. *Human-Computer Interaction, 19*, 61–84.

Saxe, R., Tenenbaum, J. B., & Carey, S. (2005). Secret agents inferences about hidden causes by 10-and 12-month-old infants. *Psychological Science, 16*, 995–1001. doi:10.1111/j.1467-9280.2005.01649.x

Scaife, M., & Van Duuren, M. (1995). Do computers have brains? What children believe about intelligent artifacts. *British Journal of Developmental Psychology, 13*, 367–377. doi:10.1111/bjdp.1995.13.issue-4

Scassellati, B., Admoni, H., & Mataric, M. (2012). Robots for use in autism research. *Annual Review of Biomedical Engineering, 14*, 275–294. doi:10.1146/annurev-bioeng-071811-150036

Smetana, J. G. (1983). Social-cognitive development: domain distinctions and coordinations. *Developmental Review, 3*, 131–147.

Smith, C. E., Blake, P. R., & Harris, P. L. (2013). I should but I won't: Why young children endorse norms of fair sharing but do not follow them. *PLoS ONE, 8*, e59510. doi:10.1371/journal.pone.0059510

Sobel, D. M., Yoachim, C. M., Gopnik, A., Meltzoff, A. N., & Blumenthal, E. J. (2007). The blicket within: Preschoolers' inferences about insides and causes. *Journal of Cognition and Development, 8*, 159–182. doi:10.1080/15248370701202356

Somanader, M. C., Saylor, M. M., & Levin, D. T. (2011). Remote control and children's understanding of robots. *Journal of Experimental Child Psychology, 109*, 239–247. doi:10.1016/j.jecp.2011.01.005

Sommerville, J. A., Hildebrand, E. A., & Crane, C. C. (2008). Experience matters: The impact of doing versus watching on infants' subsequent perception of tool-use events. *Developmental Psychology, 44*, 1249–1256. doi:10.1037/a0012296

Stanton, C. M., Kahn, P. H., Jr., Severson, R. L., Ruckert, J. H., & Gill, B. T. (2008). Robotic animals might aid in the social development of children with autism. *Human-Robot Interaction*, 271–278.

Sytsma, J., & Machery, E. (2012). The two sources of moral standing. *Review of Philosophy and Psychology, 3*, 303–324. doi:10.1007/s13164-012-0102-7

Turiel, E. (1983). *The development of social knowledge: Morality and convention.* Cambridge, UK: Cambridge University Press.

Vaish, A., Missana, M., & Tomasello, M. (2011). Three-year-old children intervene in third-party moral transgressions. *British Journal of Developmental Psychology, 29*(1), 124–130.

Weller, D., & Lagattuta, K. (2012). Helping the in-group feels better: Children's judgments and emotion attributions in response to prosocial dilemmas. *Child Development, 84*, 253–268. doi:10.1111/j.1467-8624.2012.01837.x

Wellman, H. M. (2011). Reinvigorating explanations for the study of early cognitive development. *Child Development Perspectives, 5*, 33–38. doi:10.1111/j.1750-8606.2010.00154.x

Woodward, A. L., Sommerville, J. A., Gerson, S., Henderson, A. M., & Buresh, J. (2009). The emergence of intention attribution in infancy. *Psychology of Learning and Motivation, 51*, 187–222.

Supplementary Materials

Table 1 Factor Loadings for the 7 Forced-Choice Questions.

	Moral Standing Factor	Physical Sentience Factor	Emotional Sentience Factor
	Prop of Variance Explained = 26%	Prop of Variance Explained = 19%	Prop of Variance Explained = 16%
Wrongness of Physical Harm	**.88**	.15	.10
Wrongness of Emotional Harm	**.82**	−.18	.06
Ability to Feel Emotional Pain	−.20	**.73**	.10
Consequences of Neglect	.15	**.70**	.04
Categorization	−.07	.37	**.77**
Ability to Feel Physical Pain	.06	.28	**.59**
Tactile Response	.33	.23	**.57**

Note. Factor analysis was conducted using varimax rotation.

Tables 2-8. Proportion of Explanation Types as a Function of Children's Forced Choice Responses and Condition. Proportions within each column add to 100% and bolded numbers represent modal responses within each column.

Table 2 Tactile Response: "If you Tickle AIBO, can AIBO feel it?"

	Yes		No	
	Autonomous	Controlled	Autonomous	Controlled
References to Physiological States	**74%**	25%	0%	0%
References to Mechanical Properties	11%	29%	**100%**	**100%**
Other/ Uncategorizable	15%	**46%**	0%	0%

Table 3 Ability to Feel Physical Pain: "If AIBO fell on the ground, could he get hurt?"

	Yes		No	
	Autonomous	Controlled	Autonomous	Controlled
References to Physiological States	36%	3%	0%	0%
References to Mechanical Properties	**50%**	**74%**	**100%**	**86%**
Other/ Uncategorizable	14%	23%	0%	14%

Table 4 Categorization: "Is AIBO more like a real dog or a stuffed dog?"

	Real Dog		Stuffed Dog	
	Autonomous	Controlled	Autonomous	Controlled
References to Physiological States	**68%**	**67%**	0%	12%
References to Mechanical Properties	32%	24%	**79%**	**70%**
Other/ Uncategorizable	0%	9%	21%	18%

Table 5 Ability to Feel Emotional Pain: "If someone was mean to AIBO, could AIBO get upset?"

	Yes		No	
	Autonomous	Controlled	Autonomous	Controlled
References to Desires and Emotions	**68%**	**52%**	0%	33%
References to Physiological States	5.5%	3.5%	0%	0%
References to Mechanical Properties	5.5%	3.5%	**100%**	**56%**
Other/ Uncategorizable	21%	41%	0%	11%

Table 6 Consequences of Neglect: "…do you think it's OK or not OK to leave AIBO alone in the closet for a week?"

	Not OK		OK	
	Autonomous	Controlled	Autonomous	Controlled
References to Desires and Emotions	**57%**	**34.5%**	0%	0%
References to Physiological States	20%	**34.5%**	20%	0%
References to Mechanical Properties	11.5%	24%	**80%**	**100%**
Other/ Uncategorizable	11.5%	7%	0%	0%

Table 7 Wrongness of Physical Harm: "….What if your [camp] counselor said it was OK to hit AIBO. Then do you think it would be OK, or not OK?"

	Not OK		OK	
	Autonomous	Controlled	Autonomous	Controlled
References to Moral Concern	**70%**	**33.3%**	0%	0%
References to External Consequences	18%	**33.3%**	**100%**	**83%**
Other/ Uncategorizable	12%	**33.3%**	0%	17%

Table 8 Wrongness of Emotional Harm: "….What if your [camp] counselor said it was OK to yell at AIBO. Then do you think it would be OK, or not OK?"

	Not OK		OK	
	Autonomous	Controlled	Autonomous	Controlled
References to Moral Concern	**83%**	**43%**	0%	0%
References to External Consequences	7%	39%	**70%**	**80%**
Other/ Uncategorizable	10%	18%	30%	20%

Are Prompts Provided by Electronic Books as Effective for Teaching Preschoolers a Biological Concept as Those Provided by Adults?

Gabrielle A. Strouse and Patricia A. Ganea

ABSTRACT

Research Findings: Prior research indicates that shared book reading is an effective method for teaching biological concepts to young children. Adult questioning during reading enhances children's comprehension. We investigated whether adult prompting during the reading of an electronic book enhanced children's understanding of a biological concept. Ninety-one 4-year-olds read about camouflage in 3 conditions. We varied how prompts were provided: (a) read by the book, (b) read by a researcher, or (c) given face to face by the researcher. There was an interaction between children's initial vocabulary level and condition. Children with low vocabulary scores gave fewer camouflage responses than their high-vocabulary peers, and this effect was particularly pronounced in the book-read condition. Children's executive function was also measured and discussed. *Practice or Policy*: Our findings indicate that under some circumstances electronic prompts built into touchscreen books can be as effective at supporting conceptual development as the same prompts provided by a coreading adult. However, children with low vocabulary skills may be particularly supported by adult-led prompting. We suggest that adult prompting be used to motivate children to test and revise their own biological theories. Once children have learned strategies for updating their concepts, electronic prompting may be useful for scaffolding children's transition to using the strategies when reading alone.

Technology, despite some resistance against it, is being increasingly embraced by parents for use with young children. Recent surveys indicate that 80% of U.S. children ages 2–4 have had experience using mobile devices (Rideout, 2013), and 65% of 3- to 4-year-olds in the United Kingdom have access to tablets at home (OfCom, 2014). In addition to home use, technology is also being incorporated into school curricula, as proficiency with technology is considered part of the Common Core State Standards (National Governors Association Center for Best Practices & Council of Chief State School Officers, 2010). Although technology may be less universal in preschool settings than in elementary settings, one survey of U.S. educators of children ages 0–4 found that 28% had classroom tablet computer access and 83% had classroom desktop or laptop computer access (Blackwell, Lauricella, Wartella, Robb, & Schomburg, 2013). A wide variety of educational software is available that targets preschool children, but how educational is it? In this research we are particularly interested in the role of interactive touchscreens in children's learning about the biological concept of camouflage.

Because shared reading experiences provide an opportunity for high-quality prompting, they make an excellent platform for teaching biological concepts to young children. Biological information often includes information that cannot be seen in everyday life with the naked eye. Books provide pictures and text that can give children a window into other worlds such as the animal kingdom or the inside of the body.

Learning About Biology Through Books

A growing number of studies indicate that young children can learn biological facts and concepts through shared book reading. For example, Walker, Gopnik, and Ganea (2015) found that 3-, 4-, and 5-year-old children can learn and transfer a biological causal relation (a "popple flower" causing the hiccups) encountered in a story context to a real context if the fictional context of the story is similar to the real-world story. Children 5–8 years of age can also learn a conceptually coherent explanation of natural selection from a story picture-book intervention and apply that explanation to a novel population of animals (Kelemen, Emmons, Schillaci, & Ganea, 2014). Ganea, Canfield, Simons-Ghafari, and Chou (2014) also found that 4- and 5-year-olds readily learned facts about three new animals' (cavies, oxpeckers, and handfish) behavior and environment when they were presented in a realistic context.

Exposure to books can also influence how children reason about animals at a more general level. For example, Waxman, Herrmann, Woodring, and Medin (2014) found that 5-year-olds who read an encyclopedia about bears were more likely to later adopt a biological pattern of reasoning about bears than children who read a book from the popular children's storybook series *The Berenstain Bears*. Children who read the more anthropomorphic book instead later reasoned that bears were more like humans. Similarly, Ganea and colleagues (2014) showed that 4- and 5-year-olds attribute more anthropomorphic traits to animals after being exposed to books that use anthropomorphic pictures and language than after being exposed to books that use factual language and realistic pictures of animals. In this study, the children likewise learned more facts about novel animals and their environment when exposed to these realistic books than when exposed to the anthropomorphic books. Finally, Geerdts, Van de Walle, and LoBue (2015) showed that preschoolers who were read anthropomorphized stories about butterflies and frogs used anthropomorphic language when retelling the stories, but children who heard stories with factual language did not. However, unlike in the studies by Ganea and colleagues (2014) and Waxman and colleagues (2014), children in this study did not attribute anthropomorphic qualities to other animals and explained other camouflage scenarios with equivalent language regardless of the type of book they had read. The authors hypothesized that the less extreme anthropomorphic portrayal of the animals in their books (animals had human-like faces and postures but appeared in natural environments) compared to the ones used in other stories (animals in houses wearing clothes and using furniture) may be one reason why they did not see the same generalization of anthropomorphism as the prior studies (Geerdts, 2015).

In other research, Ganea, Ma, and DeLoache (2011) showed that children could learn and transfer broader, generalizable conceptual knowledge from picture books. They found that 4-year-olds could learn to reason about the biological adaptation of camouflage from reading a realistic picture book about animals on various backgrounds. Prior to reading, children's conceptual knowledge did not allow them to generate camouflage-based reasoning about why a bird would choose a particular prey; however, after reading children not only generated camouflage-based explanations for the animals presented in the book (frogs) but also transferred that knowledge to different live animals in tanks (lizards and crabs).

Thus, shared reading experiences appear to be an excellent source for preschoolers to develop generalizable biological concepts. In this study we built on Ganea and colleagues' (2011) study by using electronic books and adding high-quality prompting to drive conceptual development about camouflage. Here we investigated whether asking children to make predictions and give explanations about camouflage, coupled with feedback, is effective in increasing children's use of appropriate camouflage-related reasoning about a predator–prey situation. In addition, we addressed whether electronic questioning is equally as supportive as questioning provided in person by varying the source of the prompts given to children: an electronic text reading itself, an adult reading text, or an adult face to face. Finally, we asked whether children's language skills or executive function moderate their learning.

The Use of High-Quality Prompts to Support Conceptual Change

Children learn best from shared reading experiences when those experiences are paired with adult scaffolding, such as with the popular program dialogic reading (Mol, Bus, de Jong, & Smeets, 2008). Research on children's story comprehension shows that adult questioning during shared reading is linked to increases in knowledge of story content. For example, when teachers prompt children for cognitively challenging talk during reading or question children about content before or after reading, children's knowledge of the content is increased (Dickinson & Smith, 1994; Morrow, 1984).

Questions that promote inferencing, or drawing connections between events in the story, seem to be especially supportive of story comprehension. Prompting children to draw inferences and modeling doing so is something that parents often do when reading with preschoolers (van Kleeck, 2008). Explicitly prompting children to make inferences could support children's understanding by focusing them on the causal chain of events in the story (Kendeou et al., 2005; Makdissi & Boisclair, 2006).

One specific type of prompt involves asking children to make *predictions* about upcoming story events. This type of prompt may act as an important support for building conceptual knowledge. Prediction involves combining existing knowledge with new information from the text to generate ideas about what might occur (Duke & Pearson, 2002). When predictions are made during reading, children who continue through the story are provided with feedback as to whether their predictions played out and have the opportunity to update their understanding. The process of integrating existing beliefs with new observations and instruction is the basis for building conceptual knowledge (Vosniadou & Ioannides, 1998). When children note inconsistencies between what they think will happen and what really happens, it provides motivation for them to update their conceptual knowledge (Carey, 2009). Thus, asking children to make predictions during shared reading, then continuing to see whether the prediction plays out, may provide motivation for children to update their relevant conceptual theories.

Although some children may generate predictions on their own while reading, Vosniadou and Ioannides (1998) noted that many children tend to think of their explanatory frameworks as facts they know rather than ideas that need to be tested. They proposed that instructional interventions for conceptual learning should make students aware of limitations in their current understanding and motivate them to reshape their beliefs. Thus, explicit prompting of children to make and explain their predictions may be needed to motivate children to recognize the inconsistences between their theories and subsequent events.

Explanation may also help encourage children to draw inferences and generalize information they have encountered. When asked to explain a mechanical event they have observed, children often use general rules that take all evidence into account (Walker, Williams, Lombrozo, & Gopnik, 2012) and focus on functional rather than superficial properties (Legare & Lombrozo, 2014; Walker, Lombrozo, Legare, & Gopnik, 2014). If asking for explanations about an event prompts children to think both generally and causally, then asking children to provide an explanation after they make a prediction may motivate them to think more deeply about their conceptual theory, potentially motivating conceptual change.

Mere motivation to update a concept, however, is not necessarily enough—to build better concepts children also need good information about *why* their prediction was incorrect (Carey, 2009). Thus, explicit *feedback* regarding children's prediction could also be an important scaffold that adults can provide.

Vosniadou (2007) argued that even very young children can make predictions and explanations about novel situations based on naïve theories built through their experiences. New experiences in a particular domain can be incorporated as children reshape their existing domain-specific theories. In this study, we gave children the opportunity to test their predictions and explanations in the biological domain using camouflage, which has been shown to be age appropriate and engaging for children ages 3–4 (Brown & Kane, 1988; Ganea et al., 2011).

It is particularly valuable to understand the nature of change in children's biological concepts. Biology is a topic that even very young children spontaneously pay attention to and engage with (Inagaki & Hatano, 2002). Naïve biology is reshaped and constructed through early life experiences, gradually growing to incorporate more advanced concepts, like natural selection. A number of individual concepts are foundational to a coherent understanding of natural selection as a process (Kelemen et al., 2014). Two such concepts are trait variation within a population and differential health and survival. It is on these individual concepts that an understanding of evolution can eventually be built. Acquiring the concept of camouflage necessarily involves understanding trait variation (color) and differential survival (invisibility to predators leads to increased survival chances). Integration of these foundational concepts may support children in later reasoning about evolutionary processes, as some have argued that evolutionary processes are counterintuitive and foundational concepts are important for building accurate evolutionary theories (Evans et al., 2010).

In the current study, we created an electronic book designed to teach 4-year-olds about the biological concept of camouflage. At three points in the book children were prompted to apply their conceptual knowledge, make a prediction about what would happen next, and give an explanation for their prediction. They were then given feedback as to whether their prediction played out the way they expected and an explanation as to why or why not.

The Use of Electronic Prompting

Although adult-led prompting has been widely studied as an educational intervention for a variety of language and comprehension skills, prompts from media sources themselves have not received much attention. Studies have shown that scripted prompts embedded into books are effective at increasing language learning when read by researchers (Ard & Beverly, 2004; Van Kleeck, Woude, & Hammett, 2006). However, to our knowledge, prompts that are read from the page have never been compared to prompts provided by a more naturalistic, face-to-face means. Because young children are highly attuned to pedagogical cues from adults to help them determine when something is instructional and should be learned or transferred (Bonawitz et al., 2011; Butler, Schmidt, Bürgel, & Tomasello, 2015; Sobel & Sommerville, 2009), we hypothesized that more naturally delivered prompts may motivate stronger conceptual development than those read from the page.

We were also interested in whether electronic books, which can adapt to children's responses, can mimic the prompting and feedback provided by adults. One prior study with video indicated that video story prompts were not as effective at supporting comprehension when provided by an on-screen actress as when provided by a coviewing parent (Strouse, O'Doherty, & Troseth, 2013). However, these video prompts were not responsive to the child, as the video continued to play regardless of the child's actions. New technologies make adaptive prompting accessible to young children. Would electronic book prompts, which are dependent on the child's tap, be a better mimic for the type of responsiveness that is present in effective adult prompting?

Finally, we were interested in whether there are certain children for whom the source of prompting would matter more than others. For example, 4- to 6-year-olds who self-generate inferences while reading have better story comprehension skills (Kendeou, Bohn-Gettler, White, & van den Broek, 2008; Tompkins, Guo, & Justice, 2013). Perhaps children with robust language and comprehension skills would be more able to self-generate inferences and revise their knowledge regardless of how the prompts are provided. Given evidence that vocabulary scores correlate with comprehension skills (Florit, Roch, Altoè, & Levorato, 2009; Kendeou et al., 2008; Sénéchal, Ouellette, & Rodney, 2006) as well as predict them (Lepola, Lynch, Laakkonen, Silvén, & Niemi,

2012), we measured vocabulary as a potential moderator. We hypothesized that children with low vocabulary skills may be particularly influenced by the source of prompting because of the relatively lower level of comprehension they may achieve without the support of effective prompting.

In addition, we hypothesized that executive function skills could play a role in children's learning from an unmediated tablet interaction versus an adult-mediated reading experience. When children are given the freedom to proceed through an electronic book on their own, children with low inhibitory control may tap through the book quickly without much thought being given to the electronic prompts. Emerging research in this area has shown that there may be some link between children's impulsive tapping behavior with tablets and their inhibitory control, particularly during instructional messages (Russo, Duncan, & Troseth, 2015). However, when an adult specifically requests information from the child, this may motivate the child to put more effort into responding to the prompts. We hypothesized that whereas children high in inhibitory control would be able to self-regulate their own experiences and learn well from all books, those low in inhibitory control may benefit more from the adult-led conditions.

Method

Participants

Participants were 91 children age 4 (mean age = 4.47 years, *SD* = 0.26; 45 females) from a large metropolitan area in Canada. Most (90%) of the participants were recruited as they visited a local science center by recruiters who introduced themselves to families as they entered the children's section of the museum. Participants varied in the exhibits they had viewed in other sections or on prior visits to the museum. In an effort to increase our sample size, we recruited an additional 10% of participants from a database of participants compiled through advertisements, local street fairs, and child care centers. An additional 21 children were not included in the analyses because the child already had knowledge of camouflage (i.e., gave camouflage-based explanations [a score of 2, described in "Pretest"] for one or both pretest questions; *n* = 10), because the child had very low English language proficiency (i.e., more than 2 *SD* below the rest of our sample; *n* = 3), because of parental interference during reading or posttest questions (*n* = 3), because the child did not want to continue (*n* = 3), because the child had a diagnosed speech production delay (*n* = 1), or because of experimenter error (*n* = 1). Children read an electronic book about camouflage during which they were prompted to make predictions, were prompted to explain their predictions, and were provided with feedback about their responses. Participants were assigned to one of three conditions in which these prompts were (a) printed in and read by the book (*book-read;* 32 children), (b) printed in the book and read by the researcher (*researcher-read;* 28 children), or (c) not printed in the book and given face to face by the researcher (*extratextual;* 31 children). Data were collected over an 8-month period. We have no reason to believe that children who participated earlier in the study were different from children participating later in the study. For the first 6 months of data collection at the science center, assignment to condition was mostly sequential, as the three books were ready for data collection at different times (researcher-read first, then book-read, then extratextual). For the final 2 months of data collection, assignment to condition was fully random. Lab-based participants were all randomly assigned to condition. Participant exclusions were spaced equally across the 8 months of data collection and across conditions. Because of the fast-paced environment at the science center we did not prescreen participants; the study was open to all 4-year-old children.

Parents selected their child's ethnicity from a list copied from the Canadian Census form or could write in their own. The sample was ethnically diverse: Participants were identified by their parents as 30.7% White (*n* = 28 children) and 26.4% South and East Asian (*n* = 24). Remaining participants were 9.8% of multiple ethnicities (*n* = 9), 4.4% Middle Eastern (*n* = 4), 3.3% Latin American (*n* = 3), 3.3% Black (*n* = 3), 2.2% Aboriginal (*n* = 2), and 4.4% other (*n* = 4). Fourteen parents declined to respond to this question. The sample was also linguistically diverse: 31% came from monolingual

Table 1. Subsample Characteristics for Participants at the Science Center and Lab Locations.

Characteristic	*Science Center* n (%)	*Lab* n (%)
Total participants included	82	9
Condition assignment		
Book-read	30 (36.6)	2 (22.2)
Researcher-read	25 (30.5)	3 (33.3)
Extratextual	27 (32.9)	4 (44.4)
Parent education (highest achieved)		
High school diploma	6 (7.3)	1 (11.1)
College diploma	12 (14.6)	2 (22.2)
Bachelor's degree	27 (32.9)	1 (11.1)
Master's degree	19 (23.2)	3 (33.3)
PhD, MD, law degree	5 (6.1)	0 (0.0)
No response	13 (15.9)	2 (22.2)
Ethnicity		
White	26 (31.7)	2 (22.2)
Aboriginal	2 (2.4)	0 (0.0)
South and East Asian	23 (28.0)	1 (11.1)
Black	2 (2.4)	1 (11.1)
Latin American	3 (3.7)	0 (0.0)
Middle Eastern	4 (4.9)	0 (0.0)
Multiple	6 (7.3)	3 (33.3)
Other	4 (4.9)	0 (0.0)
No response	12 (14.6)	2 (22.2)
Language		
Monolingual	25 (30.5)	3 (33.3)
English primary bilingual	26 (31.7)	4 (44.4)
Other primary bilingual	31 (37.8)	2 (22.2)
Reasons for exclusion		
Passing pretest	8	2
Low English proficiency/language delay	4	0
Parental interference/child not interested	5	1
Experimenter error	1	0
Total participants excluded	18	3

English homes, 33% came from multilingual homes with English as their primary language of exposure, and 36% came from homes in which English was not their primary language of exposure. Parents were generally well educated, with a median and modal response of a bachelor's degree. The breakdown of these sample demographics across recruitment methods is presented in Table 1.

Materials

Electronic Picture Book

Children viewed an electronic picture book designed to teach children about camouflage. The book was based on that used by Ganea and colleagues (2011). The book contained a title page followed by 26 pages of text. The text provided basic information about color camouflage (e.g., "The lizard is the same colour as the grass. It's hard to see the lizard so the bird will fly away") and featured three different-colored lizards each on two backgrounds (one matching, one nonmatching). After some instructional text about each lizard, the text asked the child to "Put the lizard where it will be safe from the bird!" and gave two background options. Children were then asked to make a prediction: "Do you think the bird will see the lizard there?" This was followed by a request for an explanation as to why or why not. Finally, the child was told whether the bird saw the lizard and caught it or did not see the lizard and flew away, along with a reminder about why this was so (e.g., "The lizard and sand are both orange so the bird doesn't see the lizard").

Children could turn the pages of the book by touching an arrow in the top right-hand corner. On some pages, the lizards were interactive—children could drag or tap them at the request of the book. The lizard made a very simple and brief plop when moved, so children could hear that they were successful in making the movement. No other interactive features or hotspots were included in the book.

Children in all conditions heard the same text and prompts. In the book-read condition, the declarative text and child-directed prompts were both printed in the book and read aloud on each page. Prerecorded narration by the lead researcher played when the pages were turned. In the researcher-read condition, the same text and prompts were printed on the page and read live by the researcher. In the extratextual prompts condition, the same declarative text was printed in the book and read by the researcher. Pages including child-directed prompts were removed from the book, resulting in six fewer pages (two per lizard). Instead, the researcher turned to the child and presented the prompts in a face-to-face interaction. Thus, children *heard* the same language as in the researcher-read condition but *saw* only the declarative text printed in the book. Prompts occurred in a conversation-like way, as they would if an adult were providing additional prompts when reading a typical storybook.

Test Items

Children were pretested for their knowledge of camouflage using laminated pictures of animals: a hawk, a brown rabbit on a brown background (camouflaged), a white rabbit on a brownish-red background (not camouflaged), a green caterpillar on a green leaf (camouflaged), and a red caterpillar on a green leaf (not camouflaged).

After reading, children were tested using the same laminated photograph of the hawk along with plastic aquarium tanks with aquarium rocks and four plastic replica animals: two lizards and two turtles.

Vocabulary

Children completed the National Institutes of Health (NIH) Toolbox Picture Vocabulary Test. This computerized receptive vocabulary test presents children with four pictures and prompts them to select the picture that represents a given vocabulary word. It uses computer adaptive testing to adaptively generate appropriate questions for the child based on his or her responses. Children are assigned a score using item response theory.

Executive Function

Children completed an electronic version of the day/night task used by Gerstadt, Hong, and Diamond (1994). This task was not automated. Other than the electronic format of the pictures, Gerstadt and colleagues' procedures were followed.

Questionnaires

Parents completed a questionnaire that included demographic questions and questions about their child's media use, including use of books, electronic books, videos, and other media.

Procedure

Children participated either in a workshop room at the science center (90%) or in a testing room on campus (10%). Procedures were identical. Children sat at a child-size table with the experimenter.

Pretest

Children were first shown the picture of the hawk and told, "This is a bird that eats little animals. It is very hungry and looking for a rabbit to eat." Children were then presented with photographs of two rabbits, one camouflaged and one not, and were asked, "Which rabbit do you think the bird will eat, this one or this one?" Once children selected a rabbit, they were asked, "Why will the bird eat this one?" followed by "Why won't the bird eat this rabbit?" for the other rabbit. The same procedure was followed for a pair of caterpillars with a slightly different prompt, "Which caterpillar do you think is safe from the bird, this one or this one?" followed by requests for explanations. The side of the camouflaged animal was counterbalanced. Children were excluded from the study if they made a correct choice *and* gave a camouflage-based answer for either pretest question (e.g., "It is green like the leaves"). In all 10 cases, the camouflage explanation was given to the final question, "Which caterpillar do you think is safe?"

Reading

Children were then presented with the electronic book. Children in the book-read condition were told, "On each page, the book will read itself. When you are ready, you touch here to go to the next page." The experimenter then allowed the child to proceed through the book on his or her own without intervening. This was done to mimic the reality of how children would typically experience an electronic book in the absence of adult prompting.

Children in the researcher-read and extratextual conditions were told, "I am going to read to you what the book says on each page. Then when you are ready, you touch here to go to the next page." In these conditions the experimenter intervened if the child skipped pages and turned the page back to finish reading the text (or give prompts).

Posttest

After reading, children were presented with the hawk photograph again and told, "This bird is still hungry and now it's looking for a lizard to eat." For the *placement* questions, children were presented with an orange lizard and two tanks, one with orange rocks and one with green rocks, and asked, "Where should we put the lizard so it will be safe from the bird?" The prompt was then repeated using a brown turtle with brown and black tanks. After each placement question the experimenter asked the child to explain, "Why do you think the lizard will be safe here?" For the *choice* questions, children were presented with a camouflaged lizard (black lizard in a black tank) and noncamouflaged lizard (orange lizard in a brown tank) and asked, "Which lizard will the bird eat?" This was followed by a camouflaged turtle (green/green) and noncamouflaged turtle (brown/orange). After each choice question the experimenter asked the child, "Why do you think the bird will eat this one?" Whether children were asked the placement or choice questions first was counterbalanced. The camouflaged animal alternated between being placed on the right and left.

Vocabulary Test

Children completed the computerized Toolbox Picture Vocabulary Test. After the researcher read the test instructions to the child, she gave the child the choice of (a) controlling the mouse or touching the screen to respond or (b) pointing and having the experimenter control the mouse.

Executive Function Test

Children completed the day/night task using the procedures of Gerstadt and colleagues (1994). Children saw PowerPoint slides of a sun or a moon and were trained to say "day" when the moon picture appeared and "night" when the sun picture appeared. Children were then given two to four practice trials that included researcher prompts as needed ("What do you say for this one?") and feedback ("Good job!" or a reminder of the rules). They then completed 14–16 additional test trials without feedback. Slides were turned by the experimenter. For a more detailed description of how training trials transitioned to test trials, see Gerstadt and colleagues. To be considered scorable, children needed to pass at least one of the four practice trials; scorable children's latency to respond to the slides was coded.

Coding

Children's tank selections on the posttest questions were coded as correct or incorrect by the experimenter at the time of participation. The responses were also coded from video by a second coder who was blind to condition. Reliability was κ = .91. Discrepancies were resolved by a review of the videos by the first author.

Children's explanations for their selections on the posttest questions were coded by the experimenter as containing camouflage reasoning on a scale of 0 (no mention of color or visibility), 1 (incomplete response about color or visibility), or 2 (full camouflage reasoning). For example, explanations such as "He's hungry" or "It's nice and warm" were given 0 points, as they did not mention the color of either the lizard or the background or the bird's ability to see or not see the lizard. Explanations such as "Orange stones" or "The bird will see it" were given 1 point because they used the camouflage-related concepts of color or visibility but did not provide a complete explanation for how color was important for visibility (i.e., they referenced the color of the lizard or background alone—not both—or they did not give a reason for the lizard's visibility [in this case we considered that the child may be mimicking language from the book without understanding that color was the key factor for visibility]). Explanations such as "It's the same color as the lizard [referring to the tank background]" and "It's a different color so the bird can eat it" were given 2 points. The responses were also coded from transcripts by a second coder who was blind to condition. Reliability, assessed using a weighted kappa, was .92. Discrepancies were resolved by review by the first author.

Children's accuracy on the day/night pretest trials and latency to respond on the test trials were coded from video by a coder blind to condition and hypothesis. A second blind coder scored 30% of the videos. Reliability was κ = .87 (one disagreement) for whether children passed or failed the practice trials. For the test trials, the intraclass correlation coefficient was r = .74 for children's average latency to respond on accurate trials. The first coder's latencies were used in the analysis.

Results

Book-Reading Experience

Children on average spent about 5.5 min with the book. There was a significant difference by condition in time spent with the book, with children in the book-read group spending less time (M = 5:13, SD = 1:34) than children in the adult-led groups (adult-read, M = 5:55, SD = 0:39; extratextual prompts, M = 6:00, SD = 0:47), $F(2, 84) = 4.42$, $p = .015$, $\eta_p^2 = .097$. However, there was no correlation between time spent with the book and either outcome measure.

Children in the adult-led groups heard all of the text and prompts, as they were stopped by the experimenter from turning the pages early. This was done by the experimenter keeping a finger on the very top edge of the screen so that the child's tap to the next page would not be recorded or pressing the "back" page button and repeating the prompt. Children in the book-read group were allowed to go through the book on their own. Children in this group listened on average to the entire text on 18 of the 26 pages.

Forced-Choice Tank Questions

There were no differences in children's correct tank choices for lizard questions versus turtle questions, so their choices were collapsed for a score out of 2 on each question type (placement, choice). For each repeated measures analysis of variance (ANOVA) reported here, condition was included as a between-subjects variable and question set (placement, choice) as a repeated measure. Here we specified children's tank choices as the dependent variable. There was no effect of condition, $F(2, 88) = 0.86$, $p = .428$, $\eta_p^2 = .019$. However, there was a significant main effect of question type, $F(1, 88) = 12.56$, $p = .001$, $\eta_p^2 = .125$. Children scored significantly higher on the placement questions (M = 1.63, SD = 0.68), which were more similar to the training questions asked in the book than the choice questions (M = 1.27, SD = 0.75). There was no significant correlation between tank choices and time spent reading ($r = -.01$, $r = .05$) or number of complete pages listened to (book-read group, $r = .27$, $r = .17$).

Because children's experiences while reading were different across books (e.g., children in the book-read conditions could continue past the prompts without voicing any of their thoughts out loud), we did not look for condition differences in children's responses to the prompts during reading. However, overall patterns across all children in the study may be informative for considering how children's concepts change. Children's correct lizard placement for the three lizards during reading was high across all trials (78%–79%). Correct placement of the in-book lizards was related to children's accuracy in the placement trials at posttest, weakly at first and then more strongly for the second and third in-book lizards: first, $r_s(80) = .244$, $p = .027$; second, $r_s(79) = .457$, $p < .001$; third, $r_s(80) = .416$, $p < .001$. This suggests that children who tested correct placements later in the book were better able to transfer their knowledge outside of the book context to the posttest placement trials. Correct placement of the lizards in the book was not related to children's accuracy in the choice trials, which required children to choose the nonmatching background (the opposite of what they practiced in the book).

One might expect that correct *predictions* during reading would reveal deeper conceptual understanding and would be related to children's tank choices not only in the placement trials but also in the choice trials, which required reasoning about the concept in a way that was not presented during reading. Indeed, children's correct predictions on the third but not first and second in-book lizards predicted children's correct tank choices in both the placement trials and choice trials, indicating that correct predictions made later during reading were related to children's use of camouflage at posttest: placement trials, $r_s(71) = .242$, $p = .039$; choice trials, $r_s(71) = .309$, $p = .008$.

Camouflage Reasoning

Children's highest scored response during the question set was used. If children did not answer either question in the set correctly, they were given a score of 0, as this indicated that they did not correctly use camouflage reasoning. A repeated measures ANOVA with camouflage reasoning score as the dependent variable yielded no effect of condition. However, there was a significant main effect of question type, $F(1, 88) = 16.02$, $p < .001$, $\eta_p^2 = .154$. Children scored significantly higher on the placement questions ($M = 1.25$ of 2, $SD = 0.90$) than the choice questions ($M = 0.86$ of 2, $SD = 0.94$). There was no significant correlation between camouflage reasoning and time spent reading ($r = -.02$, $r = -.20$) or number of complete pages listened to (book-read group, $r = .11$, $r < .01$). There were, however, significant correlations between children's explanations during reading and children's explanations on both placement and choice trials at posttest: placement trial first lizard, $r_s(48) = .464$, $p = .001$; placement trial second lizard, $r_s(49) = .330$, $p = .018$; placement trial third lizard, $r_s(47) = .306$, $p = .032$; choice trial first lizard, $r_s(48) = .491$, $p < .001$; choice trial second lizard, $r_s(49) = .615$, $p < .001$; choice trial third lizard, $r_s(47) = .588$, $p < .001$.

Vocabulary

To test for the effect of vocabulary, we computed a median split and categorized children as low or high scorers. This was done because we did not expect the effect of vocabulary to be constant across conditions, as would be modeled with a covariate. Instead, we hypothesized a moderation, in which vocabulary would be more predictive of success in the book-prompt condition and less influential in the adult-prompt conditions. Vocabulary level was added to the repeated measures ANOVA as a between-subjects factor with tank choice as the dependent variable. There was again no main effect of condition and a main effect of question type, $F(1, 83) = 11.15$, $p = .001$, $\eta_p^2 = .118$. There was also a main effect of vocabulary level and a Condition × Vocabulary interaction: vocabulary level, $F(1, 83) = 4.66$, $p = .034$, $\eta_p^2 = .053$; Condition × Vocabulary interaction, $F(2, 83) = 3.21$, $p = .046$, $\eta_p^2 = .072$. Children with higher vocabulary levels outscored those with lower vocabulary levels across conditions. Follow-up tests indicated that this difference was nonsignificant for the two adult-led conditions and was significant and especially pronounced for the book-read condition, $F(1, 30) = 5.14$, $p = .031$, $\eta_p^2 = .146$.

A similar pattern emerged when the camouflage reasoning score was used as the dependent variable. There was again no main effect of condition and a main effect of question type, $F(1, 83) = 11.20$, $p = .001$, $\eta_p^2 = .119$. There was a main effect of vocabulary level and a Condition × Vocabulary interaction: vocabulary level, $F(1, 83) = 10.17$, $p = .002$, $\eta_p^2 = .109$; Condition × Vocabulary interaction, $F(2, 83) = 4.88$, $p = .010$, $\eta_p^2 = .105$. Children with higher vocabulary levels gave more camouflage-based responses than those with lower vocabulary levels across conditions. Follow-up tests indicated that this difference was nonsignificant within the two adult-led conditions and was significant and especially pronounced for the book-read condition, $F(1, 30) = 26.30$, $p < .001$, $\eta_p^2 = .467$.

Executive Function

To test for the effect of executive function, we again computed a median split based on latency to correct responses in the day/night task (for those children who passed the practice trials). Executive function split was entered into the repeated measures ANOVA as a between-subjects factor (along with condition; vocabulary was removed). Tank choice was the dependent variable. There was again no main effect of condition and a main effect of question type, $F(1, 67) = 6.95$, $p = .010$, $\eta_p^2 = .094$. There was no main effect of executive function, but there was an interaction between executive function and condition, $F(2, 67) = 6.69$, $p = .002$, $\eta_p^2 = .167$. Follow-up tests indicated that the executive function split was not predictive of correct tank choices in the book-read or adult-read conditions. However, children in the extratextual condition who had slower response times on the executive function task scored worse on the tank choices, $F(1, 20) = 10.71$, $p = .004$, $\eta_p^2 = .349$. When camouflage reasoning was entered as the dependent variable, there were no significant effects except the main effect of question type, $F(1, 67) = 10.91$, $p = .002$, $\eta_p^2 = .140$.

We were also interested in whether children whose executive function measure was unscorable using Gerstadt and colleagues' (1994) procedures were different from children who passed the practice trials. Unscorable children completed two practice cards, responded incorrectly or inconsistently to four training trials, and then subsequently responded incorrectly to the first two test trials. These children could have failed for a variety of reasons, of which we propose two that are potentially primary: (a) They understood the task but had extremely low inhibitory control such that they could not inhibit the incorrect response, or (b) they had very low verbal comprehension skills and did not understand the task. In either case, the unscorable children may have been a group of children who would particularly benefit from adult scaffolding. To examine this, we entered whether children were unscorable or scorable on the day/night task as a between-subjects variable and tank choice as the dependent variable. An interaction between failure and condition emerged, $F(2, 76) = 4.07$, $p = .021$, $\eta_p^2 = .097$. There were no other significant effects. Follow-up tests indicated no difference in the adult-led conditions, but children in the book-read condition who passed the practice trials scored significantly higher on tank choice than children who failed the practice trials, $F(1, 30) = 7.54$, $p = .010$, $\eta_p^2 = .201$. When camouflage reasoning was the dependent variable, there was no significant interaction but a main effect of question set and a main effect of day/night failure: question set, $F(1, 76) = 4.42$, $p = .039$, $\eta_p^2 = .055$; day/night failure, $F(1, 76) = 8.82$, $p = .004$, $\eta_p^2 = .104$. Those who passed the practice trials had significantly higher reasoning scores (placement, $M = 1.29$, $SD = 0.90$; choice, $M = 0.93$, $SD = 0.94$) than those who failed (placement, $M = 0.57$, $SD = 0.79$; choice, $M = 0.00$, $SD = 0.00$). Reasoning scores in the group who failed the day/night task exhibited a floor effect and thus were not viable for detecting condition differences.

Discussion

In this study we investigated whether an electronic picture book with high-quality prompts was effective in increasing children's use of appropriate camouflage-related reasoning about a predator–

prey situation. Overall we found the book to be an effective tool: 74% of children across conditions used camouflage-based reasoning at the posttest compared to 2% at pretest (two children scored 1 point at pretest; children who gave 2-point responses were excluded).

We also addressed whether electronic questioning was equally as supportive as questioning provided in person through reading or face-to-face prompting. We predicted that children may particularly benefit from adult-led face-to-face scaffolding, as this condition provided the strongest social cues to pedagogy. Children may rely on social information to help them determine what information is intended to be taught and learned, and direct prompting by adults is a familiar pedagogical situation. We also predicted that children may benefit from either type of adult-led prompting over book-read prompts, as these prompts were devoid of social cues entirely. In addition, children in the book-read condition had to structure their own book-reading session—they did not have the added benefit of an adult making them stop to hear the full prompts and then look expectantly to them for answers. We found no overall differences by source of prompting in children's use of the camouflage concept to choose the correct tank at posttest or in their camouflage-based reasoning scores.

Condition differences did emerge when we looked at particular groups of children. We hypothesized that children with low language and low executive function scores may particularly benefit from adult-led face-to-face scaffolding. The results showed this to be the case. There were no effects of vocabulary level for either adult-led condition, but there was a significant difference in conceptual learning in the book-read condition. Children with high vocabularies used more camouflage-based reasoning after hearing book-read prompts than children with low vocabulary scores. One possible explanation for this effect is that children with high vocabulary scores are better comprehenders in the absence of adult prompting (Lepola et al., 2012). As such, they may exhibit high comprehension regardless of the prompting source. They may be doing a lot of prediction making, evaluating, and revising of their concepts as they read, which could act as a protective factor to allow them to learn concepts well from many sources. In contrast, children low in vocabulary may be particularly sensitive to the source of scaffolding because they are more dependent on prompting to lead them to revise their concepts. They may be reliant on the structure of adult-led scaffolding, which they are somewhat obliged to listen and respond to, compared to book-read prompts that can be clicked past with little attention.

Condition differences also emerged when we split children by their level of inhibitory control. We hypothesized that children with low inhibitory control may tap through the book quickly, ignoring prompts and potentially scoring poorly in the book-read condition, in which they had the freedom to skip pages. However, children low in executive function did not do particularly poorly in this condition. When looking at children who produced scorable results on our executive function measure, we found a difference in conceptual learning only in the extratextual condition. In this condition, children with low executive function gave fewer correct tank choices than children with higher executive function. We hypothesize that children with low executive function did particularly poorly in this condition because they were asked to wait and converse with an adult before continuing through the book. Prompts that were delivered by the adult in this condition were delivered in lieu of two pages of text, which included tapping to turn the page and visual displays. Perhaps those low in inhibitory control lost patience listening to the adult and focused their attention on their desire to continue with the story.

Finally, condition differences also emerged in relation to whether children were able to pass the practice trials of the executive function task or whether they failed and were unscorable. We proposed two main reasons for failure—very low executive function and low verbal comprehension—either of which could be reasons why children would benefit from prompting. There were no differences in tank choice in the adult-led conditions, but in the book-read condition children who were unscorable selected the correct tanks much less often than those who were scorable. Children who were unscorable performed at the floor for camouflage-related reasoning across conditions, making an exploration of condition differences untenable.

Together these findings lend support to two main arguments: (a) Electronic books accompanied by high-quality prompts can be an effective source for developing many preschoolers' biological concept of camouflage, and (b) particular sources of prompting are more effective for supporting this type of conceptual learning in some groups of children.

Electronic books may be a good choice for teaching preschoolers because children enjoy them. For example, one study found that more preschoolers showed higher levels of engagement when reading an electronic than print book than vice versa (Chiong, Ree, Takeuchi, & Erickson, 2012). A child who is uninterested in learning from traditional paper books may be excited about using a tablet to explore the same material.

Our general finding that electronic books are an effective tool for learning is in line with a recent meta-analysis. Takacs, Swart, and Bus (2014) reviewed 29 studies and found that children had equivalent comprehension of multimedia stories and traditional stories that were scaffolded by adults. They argued that multimedia features could provide scaffolding that would mimic that which an adult provides. Because our prompts were founded on principles of conceptual change that have been applied successfully in other domains, such as physics (Vosniadou & Brewer, 1992) and math (Vosniadou & Vamvakoussi, 2006), and are generally discussed as applicable across domains (Carey, 2009), we expect that our findings may hold across other scientific and nonscientific domains of learning. Encouraging children to test their theories through prediction and explanation and adapt them when necessary in a context they find motivating and engaging is a basic formula for encouraging theory updating and revision.

Our findings also offer potential caveats. In the majority of cases, the pros and cons of interactive scaffolding may balance, such that the source of prompting does not make a significant impact on learning. However, when children are at particular risk for low comprehension, the source of scaffolding may be important. Although electronic books are motivating because of their form, they may not motivate children to reflect and evaluate their conceptual theories in the same way as human interactions. This could be for a few reasons: Social cues emphasize that information is pedagogical and should be learned; there is social pressure to respond contingently and effortfully; and, at least in this study, children are required to slow down and listen to all of the prompts. Teasing apart the importance of these aspects of in-person prompting is an important area for future research.

As a final note, Duke and Pearson (2002) suggested that one good way to support comprehension is to begin with adult-led comprehension strategies and gradually release the control of those strategies to students. Electronic media can act as a supportive intermediary, reminding children to implement strategies on their own. Once children are no longer at risk for low comprehension, electronic books with built-in scaffolding may be a fun and useful educational tool for conceptual learning that children can explore on their own.

Acknowledgments

Thank you to the families who participated in the research; to the Ontario Science Centre for its invaluable collaboration and support in our data collection efforts; to Heidi Xin Yi Kuang for data coding and collection; and to Ruth Lee, Begüm Özdemir, and others in the Language and Learning Lab for helpful feedback during the design of the study.

Funding

This research was supported by Insight Development Grant No. 430-2014-00431 from the Social Sciences and Humanities Research Council, Government of Canada.

ORCID

Gabrielle A. Strouse http://orcid.org/0000-0002-6955-4955

References

Ard, L. M., & Beverly, B. L. (2004). Preschool word learning during joint book reading: Effect of adult questions and comments. *Communication Disorders Quarterly, 26*, 17–28. doi:10.1177/15257401040260010101

Blackwell, C. K., Lauricella, A. R., Wartella, E., Robb, M., & Schomburg, R. (2013). Adoption and use of technology in early education: The interplay of extrinsic barriers and teacher attitudes. *Computers & Education, 69*, 310–319. doi:10.1016/j.compedu.2013.07.024

Bonawitz, E., Shafto, P., Gweon, H., Goodman, N. D., Spelke, E., & Schulz, L. (2011). The double-edged sword of pedagogy: Instruction limits spontaneous exploration and discovery. *Cognition, 120*, 322–330. doi:10.1016/j.cognition.2010.10.001

Brown, A. L., & Kane, M. J. (1988). Preschool children can learn to transfer: Learning to learn and learning from example. *Cognitive Psychology, 20*, 493–523. doi:10.1016/0010-0285(88)90014-X

Butler, L. P., Schmidt, M. F. H., Bürgel, J., & Tomasello, M. (2015). Young children use pedagogical cues to modulate the strength of normative inferences. *British Journal of Developmental Psychology, 33*, 476–488. doi:10.1111/bjdp.12108

Carey, S. (2009). *The origin of concepts*. New York, NY: Oxford University Press.

Chiong, C., Ree, J., Takeuchi, L., & Erickson, I. (2012). *Print books vs. e-books: Comparing parent-child co-reading on print, basic, and enhanced e-book platforms*. Retrieved from the Joan Ganz Cooney Center website: http://www.joanganzcooneycenter.org/publication/quickreport-print-books-vs-e-books/

Dickenson, D. K., & Smith, M. W. (1994). Long-term effects of preschool teachers' book readings on low-income children's vocabulary and story comprehension. *Reading Research Quarterly, 29*, 104–122. doi:10.2307/747807

Duke, N. K., & Pearson, P. D. (2002). Effective practices for developing reading comprehension. In A. E. Farstrup & S. J. Samuels (Eds.), *What research has to say about reading instruction* (3rd ed., pp. 205–242). Newark, DE: International Reading Association.

Evans, E. M., Spiegel, A. N., Gram, W., Frazier, B. N., Tare, M., Thompson, S., & Diamond, J. (2010). A conceptual guide to natural history museum visitors' understanding of evolution. *Journal of Research in Science Teaching, 47*, 326–353. doi:10.1002/tea.20337

Florit, E., Roch, M., Altoè, G., & Levorato, M. C. (2009). Listening comprehension in preschoolers: The role of memory. *British Journal of Developmental Psychology, 27*, 935–951. doi:10.1348/026151008X397189

Ganea, P. A., Canfield, C. F., Simons-Ghafari, K., & Chou, T. (2014). Do cavies talk? The effect of anthropomorphic picture books on children's knowledge about animals. *Frontiers in Psychology, 5*. doi:10.3389/fpsyg.2014.00283

Ganea, P. A., Ma, L., & DeLoache, J. S. (2011). Young children's learning and transfer of biological information from picture books to real animals. *Child Development, 82*, 1421–1433. doi:10.1111/j.1467-8624.2011.01612.x

Geerdts, M. S. (2015). (Un)real animals: Anthropomorphism and early learning about animals. *Child Development Perspectives, 10*, 10–14. doi:10.1111/cdep.12153

Geerdts, M. S., Van De Walle, G. A., & LoBue, V. (2015). Learning about real animals from anthropomorphic media. *Imagination, Cognition and Personality: Consciousness in Theory, Research, and Clinical Practice*. Advanced online publication. doi:10.1177/0276236615611798

Gerstadt, C. L., Hong, Y. J., & Diamond, A. (1994). The relationship between cognition and action: Performance of children 3½-7 years old on a Stroop-like day-night test. *Cognition, 53*, 129–153. doi:10.1016/0010-0277(94)90068-X

Inagaki, K., & Hatano, G. (2002). *Young children's naive thinking about the biological world*. New York, NY: Psychology Press.

Kelemen, D., Emmons, N. A., Schillaci, R. S., & Ganea, P. A. (2014). Young children can be taught basic natural selection using a picture-storybook intervention. *Psychological Science, 25*, 893–902. doi:10.1177/0956797613516009

Kendeou, P., Bohn-Gettler, C., White, M. J., & Van Den Broek, P. (2008). Children's inference generation across different media. *Journal of Research in Reading, 31*, 259–272. doi:10.1111/j.1467-9817.2008.00370.x

Kendeou, P., Lynch, J. S., Van Den Broek, P., Espin, C. A., White, M. J., & Kremer, K. E. (2005). Developing successful readers: Building early comprehension skills through television viewing and listening. *Early Childhood Education Journal, 33*, 91–98. doi:10.1007/s10643-005-0030-6

Legare, C. H., & Lombrozo, T. (2014). Selective effects of explanation on learning during early childhood. *Journal of Experimental Child Psychology, 126*, 198–212. doi:10.1016/j.jecp.2014.03.001

Lepola, J., Lynch, J., Laakkonen, E., Silvén, M., & Niemi, P. (2012). The role of inference making and other language skills in the development of narrative listening comprehension in 4–6-year-old children. *Reading Research Quarterly, 47*, 259–282.

Makdissi, H., & Boisclair, A. (2006). Interactive reading: A context for expanding the expression of causal relations in preschoolers. *Written Language & Literacy, 9*, 177–211. doi:10.1075/wll.9.2

Mol, S. E., Bus, A. G., De Jong, M. T., & Smeets, D. J. (2008). Added value of dialogic parent-child book readings: A meta-analysis. *Early Education & Development, 19*, 7–26. doi:10.1080/10409280701838603

Morrow, L. M. (1984). Reading stories to young children: Effects of story structure and traditional questioning strategies on comprehension. *Journal of Literacy Research, 16*, 273–288. doi:10.1080/10862968409547521

National Governors Association Center for Best Practices & Council of Chief State School Officers. (2010). *Common Core State Standards.* Washington, DC: Authors.

OfCom. (2014). *Children and parents: Media use and attitudes report.* Retrieved from http://stakeholders.ofcom.org.uk/binaries/research/media-literacy/media-use-attitudes-14/Childrens_2014_Report.pdf

Rideout, V. (2013). *Zero to eight: Children's media use in America 2013.* San Francisco, CA: Common Sense Media.

Russo, C. E., Duncan, C., & Troseth, G. L. (2015, March). *iTap on a touchscreen: The role of working memory skills.* Paper presented at the biennial meeting of the Society for Research in Child Development, Philadelphia, PA.

Sénéchal, M., Ouellette, G., & Rodney, D. (2006). The misunderstood giant: On the predictive role of early vocabulary to future reading. *Handbook of Early Literacy Research, 2*, 173–182.

Sobel, D. M., & Sommerville, J. (2009). Rationales in children's causal learning from others' actions. *Cognitive Development, 24*, 70–79. doi:10.1016/j.cogdev.2008.08.003

Strouse, G. A., O'Doherty, K., & Troseth, G. L. (2013). Effective coviewing: Preschoolers' learning from video after a dialogic questioning intervention. *Developmental Psychology, 49*, 2368–2382. doi:10.1037/a0032463

Takacs, Z. K., Swart, E. K., & Bus, A. G. (2014). Can the computer replace the adult for storybook reading? A meta-analysis on the effects of multimedia stories as compared to sharing print stories with an adult. *Frontiers in Psychology, 5*. doi:10.3389/fpsyg.2014.01366

Tompkins, V., Guo, Y., & Justice, L. M. (2013). Inference generation, story comprehension, and language skills in the preschool years. *Reading and Writing, 26*, 403–429. doi:10.1007/s11145-012-9374-7

Van Kleeck, A. (2008). Providing preschool foundations for later reading comprehension: The importance of and ideas for targeting inferencing in storybook-sharing interventions. *Psychology in the Schools, 45*, 627–643. doi:10.1002/pits.v45:7

Van Kleeck, A., Woude, J. V., & Hammett, L. (2006). Fostering literal and inferential language skills in Head Start preschoolers with language impairment using scripted book-sharing discussions. *American Journal of Speech-Language Pathology, 15*, 85–95. doi:10.1044/1058-0360(2006/009)

Vosniadou, S. (2007). The cognitive-situative divide and the problem of conceptual change. *Educational Psychologist, 42*, 55–66. doi:10.1080/00461520709336918

Vosniadou, S., & Brewer, W. F. (1992). Mental models of the earth: A study of conceptual change in childhood. *Cognitive Psychology, 24*, 535–585. doi:10.1016/0010-0285(92)90018-W

Vosniadou, S., & Ioannides, C. (1998). From conceptual development to science education: A psychological point of view. *International Journal of Science Education, 20*, 1213–1230. doi:10.1080/0950069980201004

Vosniadou, S., & Vamvakoussi, X. (2006). Examining mathematics learning from a conceptual change point of view: Implications for the design of learning environments. In L. Verschaffel, F. Dochy, M. Boekaerts, & S. Vosniadou (Eds.), *Instructional psychology: Past, present and future trends. Fifteen essays in honour of Erik De Corte* (pp. 55–70). Oxford, UK: Elsevier.

Walker, C. M., Gopnik, A., & Ganea, P. A. (2015). Learning to learn from stories: Children's developing sensitivity to the causal structure of fictional worlds. *Child Development, 86*, 310–318. doi:10.1111/cdev.2015.86.issue-1

Walker, C. M., Lombrozo, T., Legare, C. H., & Gopnik, A. (2014). Explaining prompts children to privilege inductively rich properties. *Cognition, 133*, 343–357. doi:10.1016/j.cognition.2014.07.008

Walker, C. M., Williams, J. J., Lombrozo, T., & Gopnik, A. (2012). Explaining influences children's reliance on evidence and prior knowledge in causal induction. In N. Miyake, D. Peebles, & R. P. Cooper (Eds.), *Proceedings of the 34th Annual Conference of the Cognitive Science Society* (pp. 1114–1119). Austin, TX: Cognitive Science Society.

Waxman, S., Herrmann, P., Woodring, J., & Medin, D. (2014). Humans (really) are animals: Picture-book reading influences five-year-old urban children's construal of the relation between humans and non-human animals. *Frontiers in Developmental Psychology, 5*. doi:10.3389/fpsyg.2014.00172

Changing Minds With the Story of Adaptation: Strategies for Teaching Young Children About Natural Selection

Natalie Emmons, Hayley Smith, and Deborah Kelemen

ABSTRACT
Research Findings: Educational guidelines recommend a delayed, piecemeal approach to instruction on adaptation by natural selection. This approach is questionable given suggestions that older students' pervasive misunderstandings about adaptation are rooted in cognitive biases that develop early. In response to this, Kelemen et al. (2014) recently showed that young children can learn a basic yet comprehensive explanation of adaptation by natural selection from a picture storybook intervention. However, this research was conducted in a laboratory-based setting with children from middle and higher socioeconomic backgrounds. To further explore the intervention's efficacy, this investigation examined whether Kelemen et al.'s (2014, Experiment 2) findings extend to a more diverse sample of children tested in a more naturalistic setting, namely, after-school programs. After a 10-min picture storybook reading that described adaptation within a fictitious but realistic mammal species, 5- to 6- and 7- to 8-year-old children's learning of adaptation was examined. Results revealed that younger and older children benefitted from the intervention; however, older children displayed pronounced learning and generalization. *Practice or Policy*: Findings confirm that children are capable of learning complex biological ideas and that comprehensive storybook interventions are simple but powerful teaching tools. Implications for instruction on natural selection are discussed.

Understanding adaptation by natural selection is fundamental to understanding the process by which species change and diversify over time. This knowledge is more than academic: In a world where economies are increasingly fueled by medical and biotechnological responses to rapidly adapting disease pathogens, pesticide-resistant insects, and ecosystems unbalanced by climate change, a grasp of evolutionary processes is becoming prerequisite for many careers and to making informed decisions about societal and bioethical issues. It is therefore sobering to note that more than 30 years of research has demonstrated that most high school students and adults misunderstand adaptation, even after formal classroom instruction (see Gregory, 2009, for a review). The fact that misunderstandings hold among biology teachers who are expected to provide instruction on the topic serves to deepen the concern (Asghar, Wiles, & Alters, 2007; Cofré, Jiménez, Santibáñez, & Vergara, 2014; Nehm, Kim, & Sheppard, 2009; Nehm & Schonfeld, 2007; Rutledge & Warden, 2000).

What accounts for the difficulty in mastering an accurate understanding of adaptation? Findings suggest that many factors may be at play. In the United States, for example, resistance can stem in part from the perception that evolutionary concepts challenge personal religious ideologies and commitments (Evans et al., 2010; Griffith & Brem, 2004; Guliuzza, 2014; Poling & Evans, 2004). The ambiguous or misleading language that textbooks use in their descriptions of adaptation can further contribute to misunderstandings (Kampourakis, 2013). However, increasingly there is a sense that even these issues might stem from a deeper underlying factor, in particular, biases in everyday thinking. Although adaptation by natural selection is a nonrandom process, crucially, it is also entirely nonpurposive and non–goal directed. Despite this, high school students and adults tend to hold purpose-based teleological misconceptions about how adaptation occurs. Specifically, rather than understanding it as a population-based process involving differential survival and reproduction, they often reason that populations become functionally specialized through transformational events. These include, for example, ideas that ancestral individuals acquired beneficial traits via deliberate actions during their lifetimes (e.g., the idea that giraffes acquired long necks because they stretched them out while reaching for food) or because the anthropomorphized forces of "Nature" or "Evolution" transformed them in goal-directed ways (see Gregory, 2009, for a review).

Note that such incorrect notions held by older students echo the kinds of cognitive biases about the natural world present from at least early childhood (Gregory, 2009; Kelemen, 2004; Rosengren, Brem, Evans, & Sinatra, 2012; Sinatra, Brem, & Evans, 2008). These reasoning biases include tendencies to teleologically assume that natural phenomena exist in order to perform functions and that the natural world is agentive and operates in intentional, designed, and purpose-driven ways (e.g., Evans, 2000, 2001; Keil, 1989; Kelemen, 1999a, 1999b, 2012; Poling & Evans, 2002). They further include assumptions that species members share an immutable underlying reality that is responsible for their shared, invariant, and unchanging traits—an implicit belief known as *essentialism* (Emmons & Kelemen, 2015; Gelman & Rhodes, 2012; Shtulman & Schulz, 2008). Although these biases are useful as everyday reasoning heuristics, they can present difficulties when it comes to understanding counterintuitive scientific mechanisms like natural selection.

For example, recent evidence suggests that although teleological and essentialist biases may start out as separate early arising impediments to evolution learning, by at least 7–8 years of age they show signs of coalescing. Specifically, around this age, priming children's teleological intuitions that animals possess traits by virtue of their survival-relevant functions serves to deepen their essentialist commitments that species members do not vary (Emmons & Kelemen, 2015; see also Shtulman & Schulz, 2008). This is problematic because representing phenotypic variation is prerequisite to understanding adaptation as a population-based selectionist process: Within-species variation is what allows the process of differential survival and reproduction to occur. Furthermore, when young children's coalescing intuitive ideas are left unchallenged, they may become habitual and entrenched (Kelemen, 2012). In time, they can create the foundation for the commonsense theoretical ideas that contribute to high school students' and adults' incorrect beliefs about adaptation as a goal-directed transformational event within the lifetime of an individual.

In short, the presence of intuitive theoretical frameworks in early childhood raises two major concerns about current standards on teaching adaptation by natural selection: The first relates to the recommended timing of teaching about adaptation, and the second relates to the method of presentation. In the United States, current science education standards suggest that a comprehensive presentation of adaptation by natural selection should happen between Grades 8 and 12 (Achieve, Inc., 2013; American Association for the Advancement of Science, 2009; National Research Council, 2012). That is, although it is recommended that teaching about conceptual pieces of the theory occur prior to that point, guidelines propose delaying the introduction of a comprehensive explanation of adaptation that clearly and explicitly theoretically integrates facts about within-species variation, environmental context, inheritance, and differential survival and reproduction over multiple generations until junior high or high school. This delayed and piecemeal approach presumably derives from concerns about children's conceptual limitations (Metz, 1995) and lack of background biological knowledge and appears even in countries (e.g., England) that now place evolutionary teaching earlier in the curriculum (Department for Education, 2014; but see Berkman, Sandell Pacheco, & Plutzer, 2008, for uneven implementation of evolution standards).

On the positive side, this recommended component-by-component approach affords teachers latitude in terms of deciding the pacing of instruction on individual component biological facts, for example, tutoring on the relationship between access to nutritious food and health and between health and survival. It also permits gradual expertise building before requiring students to combine concepts into a multifaceted causal explanation (e.g., Kampourakis, 2013). However, from another perspective this approach is a concern: To the extent that they are rooted in inherent human cognitive biases, children's early developing intuitive theories about nature are likely to be difficult to revise or suppress once constructed. In the absence of competition from accurate theoretical alternatives that are sufficiently coherent to effectively challenge them, children's untaught intuitions can become long-term obstacles to elaborating a scientifically accurate understanding of adaptation (see Järnefelt, Canfield, & Kelemen, 2015, and Kelemen & Rosset, 2009, for more on this dual processing perspective). Prior developmental research also suggests that, theoretically speaking, young children should be conceptually capable of constructing coherent scientific theoretical alternatives when such alternatives are presented. Namely, they acquire biological factual knowledge readily (Carey, 1985; Inagaki & Hatano, 2002; Siegal & Peterson, 1999) and have robust explanatory drives and capacities for abstract theory building (Carey, 1985; Gopnik & Meltzoff, 1997; Keil, 1989; Samarapungavan & Wiers, 1997; Wellman & Gelman, 1992).

In light of these considerations, interdisciplinary research has begun to explore the viability of leveraging young children's explanatory capacities and interest in biological information to teach them about evolution. For example, recent studies have found that narrative-based approaches can promote a greater acceptance of species change and awareness of some aspects of the adaptation process (e.g., awareness that traits are passed on from one generation to the next; Browning & Hohenstein, 2013; Legare, Lane, & Evans, 2013; see also Spiegal et al., 2012, for a related museum-based approach). Drawing from these findings and others in cognitive developmental psychology, the learning sciences, and science education research, Kelemen, Emmons, Seston Schillaci, and Ganea (2014) explored whether a narrative-based approach could further yield a basic yet comprehensive understanding of adaptation by natural selection. The crucial measure was whether children could acquire a complete and cohesive understanding of adaptation without holding any misconceptions.

Kelemen et al. (2014) therefore designed a picture storybook–based intervention in which 5- to 8-year-old children listened to a custom-made factual narrative about adaptation within a novel species called the *pilosas* and then answered questions about it. The picture book causally and cohesively wove together a series of biological facts to mechanistically explain how—through the process of differential survival and reproduction—a particular trait (i.e., thinner trunks) came to predominate in the phenotypically variable population of pilosas. Specifically, following climate change–induced effects on the behavior of their insect food source, which migrated into thin underground tunnels, species members who happened to have thinner trunks were more successful at reaching their prey and thus out-survived and out-reproduced members with wider trunks over generations and time.

Results from the first of two studies revealed that both 5- to 6-year-old and 7- to 8-year-old children benefitted from the intervention (Kelemen et al., 2014). Although nearly all of the younger children and the majority of older children did not understand adaptation at pretest, following the storybook, many younger children and nearly all of the older children provided an accurate selectionist explanation of adaptation absent any misconception. Furthermore, in tests of deeper learning, children correctly generalized their understanding to explain adaptation within an entirely different novel species, and their learning endured following a delay of 3 months. Perhaps because the initial storybook described a case of rapid natural selection resulting from the rapid die-off of pilosas with wider trunks, children's selectionist explanations of adaptation centered heavily on the concept of differential survival. Thus, to further explore whether children could more fully incorporate the concept of differential reproduction into their explanations, a revised storybook that emphasized a more gradual process of adaptation occurring over many successive generations was used in the second study. After hearing this version of the book, most of the younger children and all of the older children incorporated both differential survival and reproduction into their responses when reasoning about the pilosas: Older children successfully generalized this level of understanding to a novel animal, but younger children, although largely successful at generalizing, showed a slight decrease in performance. Indeed, across both studies, older children outperformed younger children, presumably in part because of their enhanced cognitive and language abilities (Kelemen et al., 2014).

Cumulatively, these results supported the idea that young children can engage in theoretical learning about complex biological processes when presented with age-appropriate, causally cohesive mechanistic explanations. However, one limitation of Kelemen et al.'s (2014) research was that it was conducted in a laboratory setting, which offered an optimal learning environment free from distractions. Although this environment was chosen as a measure of caution given that this initial research departed significantly from accepted educational wisdom about children's learning abilities (e.g., Bransford, Brown, & Cocking, 2000; Metz, 1995; Schweingruber, Duschl, & Shouse, 2007), it did not represent a typical learning environment. A further limitation was an unintended artifact of most laboratory-based volunteer samples: Children were predominantly monolingual Caucasians from middle and higher socioeconomic backgrounds (e.g., 48% of parents in Experiment 2 had a graduate-level degree). Given that the goal of this research program is to develop an intervention with far-reaching benefits for all children, it was of interest to know the extent to which children's learning gains generalize to other populations and testing environments. The present investigation therefore sought to replicate Experiment 2 of Kelemen et al. (2014) by examining learning among children from far more diverse socioeconomic and language backgrounds and in the context of a more naturalistic learning environment. Specifically, 5- to 6- and 7- to 8-year-old children from within urban school districts that predominantly serve members of underrepresented minority groups participated in the intervention at their local after-school programs.

Method

Participants

Sixteen 5- to 6-year-old kindergarten (8 boys, 8 girls, *M* age = 6 years, 2 months, *SD* = 6 months) and sixteen 7- to 8-year-old second-grade children (7 boys, 9 girls, *M* age = 8 years, 3 months, *SD* = 3 months) were recruited from three Boston-area after-school programs. Children were racially and ethnically diverse (50% Hispanic, 22% multiracial, 16% Caucasian, 9% African American, 3% unknown). Of the children whose parents reported details about their child's language(s) (78%), 60% were identified as speaking a second language: Most of those children (87%) were bilingual Spanish–English speakers.

Information about household income and parent education was collected to provide indices of socioeconomic status: The average annual household income was between $50,000 and $60,000 (*SD* = $50,000; 50% response rate), and the average level of parent education included some college (it ranged from less than seventh grade to a graduate-level degree; 53% response rate). This indicated that children came from diverse economic and educational backgrounds. Finally, parent responses to a question about the type of explanation they would provide their child about adaptation confirmed that children were not from backgrounds in which they were likely to receive extensive or accurate explanations of natural selection (94% of parents did not accurately describe natural selection; 56% response rate).[1] This is consistent with patterns found in the original set of studies by Kelemen et al. (2014) despite parents in those studies having higher levels of education on average.

[1]Parents were asked to write down what they would tell their child if their child asked them the following: "How did the giraffe get its long neck?" Their open-ended responses were classified using the same criteria used to classify children's responses with the exception that knowledge of the facts was not evaluated and thus not included in the criteria.

Materials and Procedure

Study Environment

As in Kelemen et al. (2014), trained experimenters individually tested children. However, all testing procedures occurred at the child's after-school program rather than in a controlled laboratory environment. Two of the three after-school programs were located within a public school. In the original study, testing took place in a quiet, calm, bare-walled laboratory testing room. By contrast, children in the present study were tested in either an unused classroom or a space that closely resembled a typical classroom in that it contained visual media on the walls and learning materials (e.g., books, games, and toys) on shelves and other surfaces. In addition, moderate levels of background noise were created by nearby activities or groups of children passing through the halls. Given that this setting was generally busier and had more distractions, it provided a stronger test of the efficacy of the storybook intervention.

Storybook

A custom-made picture storybook was used because of the absence of commercially available picture books that accurately and comprehensively explain the logic of adaptation by natural selection. The book used was the same as that used in Experiment 2 of Kelemen et al. (2014). In general, a picture book format was implemented because it has numerous benefits: Children are naturally interested in picture books, and storybooks invite a beneficial joint attentional learning context (e.g., Moore & Dunham, 1995; Tomasello & Farrar, 1986). Furthermore, by presenting a verbal narrative enriched by visual images, a picture book reduces children's cognitive load (Mayer & Moreno, 2003) while simultaneously supporting the paced sequential unfolding of a multifaceted causal explanation that fluidly incorporates new conceptual elements on a page-by-page basis (see also Kelemen et al., 2014).

The book had realistic color drawings of a fictional species in its habitat. All illustrations were attractive yet deliberately unembellished by unnecessary detail, extraneous features, or garish color to avoid distracting from the causal explanation presented in the narrative text and images (DeLoache, 2004; Tare, Chiong, Ganea, & DeLoache, 2010). Across 12 pages, the book explained how the fictitious pilosas species went from having mostly wider trunks in the past to having mostly thinner trunks in the present because of the process of differential survival and reproduction. Seven key biological concepts were causally woven together in the text: (a) the inherent variation of traits within a population, (b) habitat and food source change as a result of climate change, (c) differential health and survival due to differential access to food, (d) differential reproduction due to differential health, (e) the reliable transmission of heritable physical traits across generations, (f) the stability and constancy of inherited traits over the life span, (g) trait frequency changes over multiple generations.

Text was simply worded and deliberately devoid of any intentional or teleological language that might lead to misinterpretation or a misconception (see Emmons & Kelemen, 2015). For example, the notion that the variable trunk widths present in the population performed different survival-related functions after the climate changed was initially introduced in the following nonteleological, nonintentional way across several storybook pages:

> The small number of pilosas with thinner trunks could eat lots of milli bugs because their trunks could fit all the way to the bottom of the holes where most of the milli bugs lived. Pilosas with wider trunks had trouble getting food. They could only fit the tips of their trunks into the holes. Some pilosas with wider trunks got to eat when they found milli bugs that were moving about near the top. But other pilosas with wider trunks did not eat anything at all. The small number of pilosas with thinner trunks were strong and healthy because of all the bugs they were eating.

The gradualness of adaptation was depicted across several pages showing that, over successive generations, pilosas with the disadvantageous trait (wider trunks) slowly diminished in number because of their reduced access to food and thus reduced survival and reproduction rates. Meanwhile, pilosas with the more advantageous trait (thinner trunks) gradually increased because of their enhanced survival and reproduction rates. The text further highlighted the concept of trait constancy because children are known to accept physical transformations over the life span as a function of inevitable growth (Hermann, French, DeHart, & Rosengren, 2013; Rosengren, Gelman, Kalish, & McCormick, 1991). By making this concept explicit, the text was intended to reduce the likelihood that children might incorrectly teleologically reason that beneficial traits can be transformationally acquired over the life span in response to need. The storybook reading took about 10 min.

Assessments

Children's understanding of natural selection was assessed a total of three times: once with a novel species before they heard the storybook (pretest) and twice following the storybook reading in two posttests. A comprehension posttest evaluated their understanding of adaptation within the pilosa species, and a generalization posttest examined their ability to apply the logic of natural selection to another unfamiliar novel species that had undergone adaptation. Materials used in the pretest and generalization posttest were counterbalanced. Because of variable and unavoidable testing practicalities at different after-school programs, some children completed the pre- and posttests on one day (10 children) and some completed the pretest on one day and both posttests on another day (22 children). It is important to note that the storybook reading always immediately preceded the comprehension posttest. Unlike in Kelemen et al. (2014, Experiment 2), there was no 10-min break between the comprehension and generalization posttests for any children because of time constraints. All assessments involved realistically drawn but visually distinct novel mammal species that had undergone adaptation on a trait relevant to gaining access to food (necks, legs, facial parts). High structural alignment was maintained across assessments because children's ability to abstract concepts across examples is facilitated by deep structural similarities (Brown & Kane, 1988; Gentner, Loewenstein, & Thompson, 2003).

The questions used in each assessment were identical to Kelemen et al. (2014, Experiment 2). For each test, children were first asked a fixed set of six closed-ended questions to probe whether they grasped the isolated facts prerequisite to supporting an understanding of natural selection (see Table 1). Before any questions were asked, children were given information about and shown images of the past and present populations and habitats of the respective species. While viewing the habitat images and two species members—one with the advantageous trait and one with the disadvantageous trait—children answered four of the closed-ended questions (i.e., two questions about differential survival and two about differential reproduction). To succeed on these questions, children had to consider the past and present trait frequency distributions (e.g., the distribution of pilosas with wider and thinner trunks) and the location of the species' food source in the past and present (e.g., where the insect food source lived). Children also had to infer that the trait of interest was relevant to gaining access to food, which was never explicitly stated during any assessment. The two other closed-ended questions probed children's understanding of inheritance and trait constancy and were not directly tied to information about the past and present populations and habitats. Children were asked to justify their responses to all closed-ended questions to see whether their reasoning was correct or incorrect, and this justification—rather than their initial forced-choice response—determined whether they were credited with knowing the concept. It is important to note that corrective feedback was never provided at any point in the assessment.

Following the isolated fact questions, children were asked four open-ended questions to examine their ability to self-generate an explanation of adaptation by natural selection (see Table 2). Children answered these questions while viewing the past and present populations. To encourage them to elaborate on their reasoning, children received prompts to expand their answers when self-generating explanations to the open-ended questions (see Table 2). Over the course of testing, all but one child interacted with two experimenters. This experimental feature was introduced in Kelemen et al. (2014, Experiment 2) to discourage children from shortcutting their answers when asked similar forms of a question repeatedly (over the course of three structurally aligned assessments) by the same person. All but six children were introduced to the second experimenter at the point of the storybook reading: The remaining children interacted with a new experimenter at the point of the generalization posttest.

Table 1. Closed-Ended Isolated Fact Questions With Sample Justifications.

Concept	*Question*	*Accurate Justification*	*Inaccurate Justification*
Differential survival	Nowadays, will a *wilkie with shorter legs* probably be healthy and live for a long time? Why/why not?	No, because they can't reach up to the yellow berries.	No, I don't know. I'm not sure.
	Nowadays, will a *rudoo with a longer neck* probably be healthy and live for a long time? Why/why not?	Yeah, because it has a longer neck and it can reach food that's higher.	No, because it's an adult, and when you're an adult, you have to die.
Differential reproduction	Nowadays, will a *rudoo with a shorter neck* probably have lots of children? Why/why not?	No, because the fruit is growing on top and they they're too small to reach it.	No, um because they might be too old to have children.
	Nowadays, will a *wilkie with longer legs* probably have lots of children? Why/why not?	Yeah, 'cause it has a lot of energy inside it.	Yeah, because it's like an adult, so it might have babies, 'cause adults always have babies.
Trait knowledge: Inheritance	These grown-up *wilkies* both have *shorter legs*. If these two *wilkies* had a child, what kind of *legs [longer or shorter]* would their child probably have? Why?	Shorter, 'cause when it is born it has the same legs as its parents.	Shorter, because it's little.
Trait knowledge: Trait constancy	See this young *rudoo*. It was born with *a shorter neck*. When this *rudoo* grows up to be an adult, what kind of *neck* will it have *[longer or shorter]*? Why?	Shorter, because ... it's just gonna be them [the rudoo] when they're like bigger so they're not gonna change that much ...	A longer neck, 'cause when you grow, you grow bigger.

Note. Italicized information differed depending on the animal species under consideration. For the differential survival and reproduction questions, the presentation of advantaged and disadvantaged animals was counterbalanced.

Table 2. Open-Ended Questions.

Many hundreds of years ago most of the grown-up *pilosas* had *wider trunks* but now most of the grown-up *pilosas* have *thinner trunks*. How do you think that happened?
What happened to *pilosas* with *thinner trunks?* Why?
What happened next after ...? [E repeats P's response to previous question] Why?
What happened next after ...? [E repeats P's response to previous question] Why?
What happened to *pilosas* with *wider trunks?* Why?
What happened next after ...? [E repeats P's response to previous question] Why?
What happened next after ...? [E repeats P's response to previous question] Why?
Did it take a short time or a long time for *pilosas* to go from having mostly *wider trunks* in the past to having mostly *thinner trunks* now? Why?

Note. Italicized information differed depending on the animal species under consideration. Question orders about advantaged and disadvantaged animals were counterbalanced. E = experimenter; P = participant.

Coding

As in Kelemen et al. (2014, Experiment 2), children's responses to all closed- and open-ended questions were coded based on a conceptual checklist and conservative coding rubric to determine their overall level of understanding of natural selection at each test time (see Table 3 for details and sample responses). This type of classification scheme looks at the whole of each child's theoretical understanding and permits examining individual learning of the complete causal mechanism of adaptation. By contrast, other approaches have explored children's learning about individual evolutionary concepts in the absence of examining each child's complete theoretical framework (e.g., Browning & Hohenstein, 2013; Legare et al., 2013). In these less conservative approaches, children were credited for understanding specific evolutionary concepts (e.g., that traits are inherited) even if they simultaneously displayed a misconception about evolution. Although such approaches are useful for evaluating learning about individual component facts, they do not provide a complete picture of each child's theoretical understanding of adaptation. The goal of the present coding

Table 3. Conceptual Checklist of NS Understanding and Sample Partial Open-Ended Responses.

Level and Checklist	*Partial Open-Ended Responses Following One or More of the Four Open-Ended Questions*[a]
Level 0: No isolated facts[b] Does not meet criteria for isolated facts	Not applicable
Level 1: Isolated facts but no NS understanding Meets criteria for isolated facts,[b] but one or more of the following is present: (a) misconception, (b) no mention of differential survival advantage, (c) inaccurate mention of differential survival or reproduction	*Example of a misconception:* E: . . . How do you think that happened? P: 'Cause they're grown up. See, see how it had short body [points to animals in past population] and see some of these have bigger bodies [points to animals in present population]. And look it's all grown up. *Example of inaccurate mention of differential survival:* E: So, what happened to pilosas with wider trunks? P: They stayed alive for a lot of days. E: And why is that? P: Because they can reach the food.
Level 2: Foundation for NS understanding All of the following are present: (a) meets criteria for isolated facts, (b) no misconception, (c) accurate mention of differential survival	*Example of correct mention of differential survival, but no mention of differential reproduction:* E. What happened to pilosas with wider trunks? P: They died. E: Why's that? P: Because they didn't eat. They like ate only one each day. E: What happened to pilosas with thinner trunks? P: They keep, they had, they lived a long time. E: Why's that? P: Because they ate a lot of it, like 11 each day. E: So what happened after they lived a long time? P: They died. E: Why's that? P: Because they got old and old and weak.

(*Continued*)

Table 3. (Continued).

Level and Checklist	*Partial Open-Ended Responses Following One or More of the Four Open-Ended Questions*[a]
Level 3: NS understanding in one generation All of the following are present: (a) meets criteria for isolated facts, (b) no misconception, (c) accurate mention of differential survival, (d) accurate mention of differential reproduction in one generation	*Example of correct mention of differential survival and differential reproduction in one generation:* E: What happened to pilosas with wider trunks? P: Then they had one baby or none because they died. E: And why did they only have one baby or no babies because they might've died? P: Um because they couldn't reach the milli bugs when it got really hot. It was all the way, they were all the way down under the ground. E: So, what happened to pilosas with thinner trunks? P: They were really healthy and got two children. E: Why is that? P: Because they were healthier and they ate a lot more milli bugs.
Level 4: NS understanding for multiple generations All of the following are present: (a) meets criteria for isolated facts, (b) no misconception, (c) accurate mention of differential survival, (d) accurate mention of differential reproduction in one generation, (e) accurate mention of differential reproduction in multiple generations	*Example of correct mention of differential survival and differential reproduction in multiple generations:* E: . . . How do you think that happened? P: When the um, the ones that had thinner, the grown-ups that had thinner ones got to eat a lot and they had a lot of children so the children had a lot of, were a lot healthy and had a lot of energy so they had a lot of children. But the ones that um have um thicker trunks they um, they couldn't eat so much so they only could have one baby so they could um, so the ones that have less, the ones that have less babies are then um, get one more baby again and then they'll just die, but the ones with um, thinner ones, thinner trunks are gonna live longer.

Note. NS = natural selection; E = experimenter; P = participant.
[a]The initial open-ended question, "How do you think that (population change) happened?" was followed by subsequent requests for elaboration (see Table 2). Sample responses reported here have been edited for length such that only partial responses, often in connection to requests for elaboration, are shown to illustrate specific concepts that were coded as part of the conceptual checklist; however, these partial responses do not reflect any one child's entire open-ended response. [b]"Meets criteria for isolated facts" was defined as correctly answering and justifying five out of six of the closed-ended questions.

scheme was to evaluate the cohesiveness, completeness, sophistication, and accuracy of children's understanding of adaptation at each test time.

For each assessment, children's understanding was classified into one of five hierarchical levels. Children's understanding was categorized as Level 0, "no isolated facts," when responses to the closed-ended isolated fact questions demonstrated insufficient knowledge of the requisite facts needed to support an understanding of natural selection regardless of responses to open-ended questions. Children who did not respond correctly to at least five out of six of the closed-ended questions fell into this category. Understanding was categorized as Level 1, "isolated facts but no natural selection understanding," when it met the minimum criteria for factual understanding (five or more of the closed-ended questions) but did not reveal a correct selectionist-based understanding of adaptation by natural selection in the open-ended questions because of a lack of knowledge about the population-based process or as a result of stating a misconception. The three remaining levels (Levels 2–4) were reserved for responses that met the minimum criteria for factual understanding and also contained an accurate population-based selectionist description of adaptation absent any misconception; however, the degree of sophistication differed for each level.

In Level 2, "foundation for natural selection understanding," children's responses focused on adaptation resulting from differential survival, that is, the concept that species members with disadvantageous traits died whereas those with advantageous traits survived. In Level 3, "natural selection understanding in one generation," children's responses explained adaptation in terms of both differential survival and differential reproduction, but ideas about the relative reproductive success of species members with advantageous traits were limited to considerations of the first generation after the climate change and its immediate descendants. Finally, in Level 4, "natural selection understanding for multiple generations," children's responses were expanded such that they explicitly acknowledged that adaptation occurs over multiple generations. It is crucial to note that a Level 2 or higher categorization was assigned only if children gave no signs of holding transformationist misconceptions that individuals acquire advantageous traits within their lifetimes at any point in the assessment. Interrater reliability between two coders was excellent ($\kappa = .97$).

Results

As in Kelemen et al. (2014), data were analyzed using repeated measures ordinal logistic regressions. These regressions examined how the distribution of children across the five hierarchical levels of natural selection understanding changed across the three assessment times (i.e., pretest, comprehension, and generalization). Odds ratio (OR) statistics from these analyses further indicated the magnitude of change in the odds that children's understanding of natural selection improved by one or more levels between two specific assessment times.

Younger Children

Repeated measures ordinal logistic regressions revealed that the storybook intervention induced learning, Wald $\chi^2(2) = 12.59$, $p < .01$, N = 48 (see Figure 1). Given younger children's starting levels of understanding at pretest, their odds of being in a higher level of natural selection understanding at comprehension increased 22-fold (OR = 22.57, $p < .001$, 95% confidence interval [CI] [3.94, 129.32]): At pretest, 88% of children lacked sufficient knowledge of the individual biological facts. After hearing the storybook, 69% of children had acquired these isolated facts and 50% also displayed a Level 2 or higher population-based understanding. Indeed, 44% of children incorporated not only differential survival but also differential reproduction into their selectionist explanations such that they were in Levels 3 or 4. At generalization, 56% of children still demonstrated knowledge of isolated facts. However, reflecting the difficulties of generalization (Brown, Kane, & Long, 1989; Gentner, 1989), there was a significant 3-fold decrease in children's odds of being in a higher level of

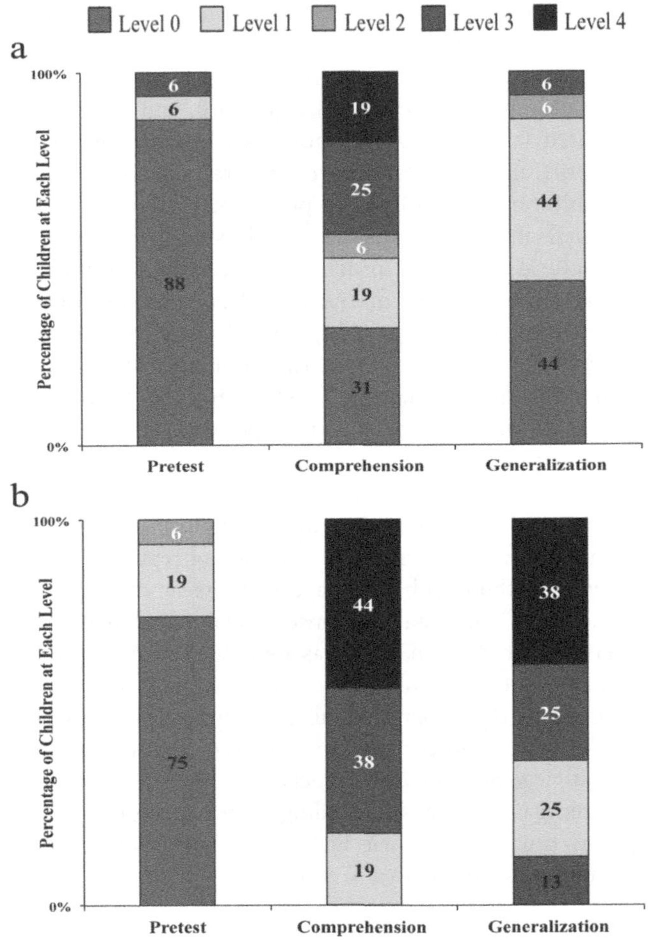

Figure 1. Percentages of (a) younger and (b) older children classified into the five levels of natural selection understanding as a function of assessment. Because of rounding, percentages do not always add up to 100%. Level 0 = no isolated facts; Level 1 = isolated facts but no natural selection understanding; Level 2 = foundation for natural selection understanding; Level 3 = natural selection understanding in one generation; Level 4 = natural selection understanding for multiple generations.

natural selection understanding between the comprehension and generalization posttests (OR = 0.30, $p = .02$, 95% CI [0.11, 0.81]).

Older Children

Repeated measures ordinal logistic regressions revealed that the revised storybook also induced learning in older children, Wald $\chi^2(2) = 31.41$, $p < .001$, N = 48 (see Figure 1). Older children's odds of being in a higher level of natural selection understanding increased a substantial 66-fold from pretest to comprehension (OR = 66.36, $p < .001$, 95% CI [15.29, 287.98]). Among older children, 75% lacked sufficient knowledge of the individual facts at pretest and another 19% had the isolated facts but had not integrated this knowledge into a population-based explanation. After hearing the storybook, 100% of children had the individual facts and 81% had a selectionist understanding of adaptation that incorporated both differential survival and reproduction. Note that 44% of children displayed the highest level of understanding (Level 4) by describing natural selection occurring over multiple generations. Children also successfully applied what they learned from the storybook to a novel animal, demonstrating a nonsignificant change in their odds of being in a higher level of natural selection understanding from comprehension to generalization ($p = .09$).

Discussion

This investigation extends earlier laboratory-based research by showing that young children from a range of socioeconomic and often bilingual backgrounds can understand and apply the basic mechanism of within-species adaptation by natural selection when learning takes place in a more distracting, naturalistic setting. Note that, as in Kelemen et al. (2014), older children outperformed younger children. However, in many respects, older children's performance in the present study was even more striking than that found in Kelemen et al. (2014). This is because in contrast to the 7- to 8-year-olds in the original research, the current sample of older children entered the study displaying far less biological knowledge. Indeed, their starting level of factual knowledge at pretest was more similar to that of the younger kindergarten-age children in the present study than the second graders in the original laboratory-based sample (Kelemen et al., 2014, Experiment 2). Like their younger counterparts then, before hearing the storybook, these older children did not initially display the prerequisite biological knowledge that would make them well prepared to understand the logic of adaptation. Despite this, after engaging with the storybook once, the majority of them not only achieved the two most sophisticated levels of natural selection understanding at the comprehension posttest but also demonstrated deep learning by generalizing what they had learned to explain population changes within a different novel animal.

By comparison, the concentration of kindergarten children's gains was in their acquisition and generalization of the isolated biological facts rather than in their deep coherent learning of the population-based selectionist theory of natural selection. However, 51% of younger children nevertheless displayed a coherent understanding of the mechanism of adaptation during the comprehension test despite their difficulty abstracting the theory and generalizing it to a novel animal. Furthermore, when younger children gave a correct population-based explanation of natural selection, they generally provided a more sophisticated account that referenced both differential survival and reproduction (Level 3) or went further by incorporating the idea that the process occurs over multiple generations (Level 4). Thus, although younger children did not demonstrate the same learning gains as older children, a substantial proportion were able to grasp the selectionist logic and articulate the multistep causal process outlined in the storybook.

Older children's marked abilities to learn and generalize the selectionist logic of natural selection both in the present study and in Kelemen et al. (2014) raise questions about the source of their enhanced learning. More developed cognitive and linguistic abilities—including improved working memory, attention, and expressive language—presumably contributed to their stronger performance (see also Kelemen et al., 2014). However, there is an additional explanation for older children's more pronounced grasp of the selectionist mechanism. Prior research has shown that 7- to 8-year-old children are better able to represent a key evolutionary concept—within-species biological variation—than 5- to 6-year-olds (Emmons & Kelemen, 2015; see also Legare et al., 2013). Within-species variation is the condition that allows the selectionist process of differential survival and reproduction to occur. Given this, children who can inhibit their essentialist tendencies such that they can represent variation should be in a better position to learn the logic of natural selection and avoid the transformationist misconceptions that contributed to many of the younger children's failures and that are widely observed among older students (Gregory, 2009). In light of this tentative interpretation, future research should seek to directly explore the relation between children's understanding of within-species variation and their abilities to learn natural selection. At least among adults, some research has found a connection between variation and adaptation understanding (Shtulman & Schulz, 2008).

Based on the developmental patterns in the present findings, one interpretation is that the storybook intervention would be most effective when introduced in second grade. Although that may be true, it may also be the case that an earlier, and possibly repeated, introduction to the storybook may help to support younger children's grasp of biological facts that have the potential to aid not only in their learning about natural selection but also in their ability to more deeply understand living things in a general sense, including understanding the biological processes they undergo (e.g., genetic inheritance). Consistent with this possibility, findings from a recent classroom intervention in which children heard two custom-made storybooks about adaptation and took part in a structured classroom discussion about their contents revealed that kindergartners' long-term factual and theoretical knowledge about natural selection was enhanced through repeated instruction (Emmons, Lees, & Kelemen, 2016). Nevertheless, and in line with the superior gains found among older children in the present investigation, second graders still outperformed kindergartners in the classroom intervention. In particular, they were far more successful in engaging in far generalization by abstracting the logic of natural selection and applying it to vastly different cases, namely, predator–prey contexts.

The present findings alongside others (Emmons et al., 2016; Kelemen et al., 2014) suggest promising avenues of evolution instruction for young children. However, it is important to note that the picture storybook intervention used here targeted microevolutionary understanding and thus is not viewed as a panacea for all of the challenges faced in learning about an array of evolutionary concepts that include common descent and speciation. More work is needed to explore whether there are far-reaching benefits of learning about adaptation early on and, more specifically, whether early learning about adaptation can serve to scaffold later learning of macroevolutionary concepts.

Implications for Education and Practitioners

Thirty years of research has shown that older students, adults, and educators have pervasive misconceptions about natural selection (Gregory, 2009). These misconceptions appear to be rooted in early emerging cognitive biases to essentialize species and teleologically explain the natural world (e.g., Kelemen, 2012; Rosengren et al., 2012) that when left unchallenged can become habitual and entrenched (Kelemen et al., 2014). Results from the current study combined with those from Kelemen et al. (2014) challenge conventional educational wisdom that young children are largely incapable of understanding causally complex abstract ideas. They therefore also challenge educational guidelines recommending that instruction should focus on gradual, piecemeal expertise building before presenting information on complex theories as a comprehensive whole. Instead, what the present findings support is that more comprehensive theoretical content can be introduced earlier in the curriculum as part of a long-term strategy that involves spaced, progressive revisitations of concepts to improve scientific literacy. By virtue of starting earlier, interventions targeting young children might help to increase the chances that a scientifically accurate understanding is acquired and maintained longer term (Emmons & Kelemen, 2015; Kelemen et al., 2014). With this in mind, how then might educators implement strategies to successfully teach counterintuitive biological concepts to young learners?

Children's abilities to extract abstract theoretical content from educational materials should not be underestimated. Children have natural drives to understand how the world works and possess robust theory-building capacities (Carey, 1985; Gopnik & Meltzoff, 1997; Keil, 1989; Wellman & Gelman, 1992). They are also naturally interested in animals and the biological world (e.g., Kelemen, Callanan, Casler, & Pérez-Granados, 2005; LoBue, Bloom Pickard, Sherman, Axford, & DeLoache, 2013). Educators can leverage these factors when introducing biological content. In particular, they can aim to provide mechanistically detailed scientific explanations that build both young children's factual knowledge and their theoretical understanding. As shown in the present study, children are capable of extracting biological factual information when it is embedded within a more extensive causally cohesive theoretical explanation. Thus, it may be entirely unnecessary—and perhaps even less ideal—to provide tutoring on individual component facts out of context of a larger theoretical framework. It is important to note that the explanations provided should be simple but also cohesive and complete to avoid explanatory gaps that could be subject to reinterpretation by children's intuitive and often incorrect causal frameworks.

On a related note, educators should be mindful of the explanatory biases children possess and avoid language that can perpetuate an incorrect understanding of biological processes such as natural selection. For example, when describing how species change, it seems advisable to avoid language that could be misconstrued as suggesting that change occurs at the individual level (e.g., "giraffes' necks grew longer over time") or is driven by need (e.g., "giraffes' necks changed because they needed to reach high leaves"). Although commonly used, these expressions were deliberately avoided in the storybook because of their ambiguity and connection to older students' transformationist misconceptions about adaptation (Emmons & Kelemen, 2015; Gregory, 2009; but see Evans, Legare, & Rosengren, 2011; Legare et al., 2013; Zohar & Ginossar, 1998). Instead, a focus on within-species variation and the idea that change occurs at a population, rather than individual, level may help to limit children's misunderstandings about adaptation. We recognize that professional development support may be required to facilitate this kind of shift in practice.

Educators may also consider developing a curriculum around scientifically sound narrative-based materials such as the picture storybook used here. As demonstrated by the present results, image-supported narrative explanations can go a long way toward facilitating understanding of complex material. Children like stories, and stories can provide a pragmatically useful entry point for active scientific inquiry guided by elementary school teachers who might otherwise feel uncomfortable or unfamiliar with the specific details of certain scientific content. However, educators should exercise care when selecting narrative materials to utilize in the classroom given that many may contain the very misconceptions that teachers are working against yet may themselves nonreflectively hold and inadvertently convey in their teaching practices (Ansberry & Morgan, 2010). Finally, although narrative-based materials can be simple but powerful learning tools, incorporating additional active learning activities may help to promote deeper understanding and enduring learning. Putting aside discussion of the potential benefits of supplementary hands-on learning opportunities, simply encouraging young children to self-explain what they have learned, is likely to facilitate deeper processing and abstraction of information (e.g., Ainsworth & Loizou, 2003; Chi, De Leeuw, Chiu, & Lavancher, 1994). Although additional research is needed, opportunities for self-explanation—like those offered by the present assessments—may be a central part of children's learning and retention of complex theoretical ideas.

Conclusion

In sum, the present study demonstrates that young children from a range of backgrounds are capable of acquiring biological facts and a basic comprehensive understanding of natural selection when learning and evaluation occurs in a school-like environment. These results illustrate the potential of early education interventions that derive from synthesizing and implementing findings from cognitive development, the learning sciences, and science education research. They also have numerous implications for biology education specifically. Our hope is that these findings will serve to foster interdisciplinary discussions about the best ways to promote long-term understanding of counterintuitive ideas and further the development of progressive theory-based education tools targeting learning from earlier ages.

Acknowledgments

The authors would like to thank the families and after-school programs that participated in this research as well as Sarah Khalifa, Kristin Lees, Laura Nelson, Mikayla Scaduto, Becca Seston Schillaci, and other members of the Child Cognition Lab.

Funding

The research in this article was supported by National Science Foundation Project Grant 1007984 awarded to Deb Kelemen. The views expressed in this article are those of the authors.

References

Achieve, Inc. (2013). *Next generation science standards*. Retrieved from http://www.nextgenscience.org/search-standards

Ainsworth, S., & Loizou, A. T. (2003). The effects of self-explaining when learning with text or diagrams. *Cognitive Science, 27*(4), 669–681. doi:10.1207/s15516709cog2704_5

American Association for the Advancement of Science. (2009). *Benchmarks for science literacy*. New York, NY: Oxford University Press.

Ansberry, K. R., & Morgan, E. R. (2010). *Picture-perfect science lessons: Using children's books to guide inquiry: Grades 3-6*. Arlington, VA: National Science Teachers Association Press.

Asghar, A., Wiles, J. R., & Alters, B. (2007). Canadian pre-service elementary teachers' conceptions of biological evolution and evolution education. *McGill Journal of Education, 42*(2), 189–210.

Berkman, M. B., Sandell Pacheco, J., & Plutzer, E. (2008). Evolution and creationism in America's classrooms: A national portrait. *PLoS Biology, 6*(5), e124. doi:10.1371/journal.pbio.0060124

Bransford, J. D., Brown, A. L., & Cocking, R. R. (2000). *How people learn: Brain, mind, experience, and school.* Washington, DC: National Academies Press.

Brown, A. L., & Kane, M. J. (1988). Preschool children can learn to transfer: Learning to learn and learning from example. *Cognitive Psychology, 20*(4), 493–523. doi:10.1016/0010-0285(88)90014-X

Brown, A. L., Kane, M. J., & Long, C. (1989). Analogical transfer in young children: Analogies as tools for communication and exposition. *Applied Cognitive Psychology, 3*, 275–293. doi:10.1002/(ISSN)1099-0720

Browning, E., & Hohenstein, J. (2013). The use of narrative to promote primary school children's understanding of evolution. *Education 3-13: International Journal of Primary Elementary and Early Years Education, 2*, 1–18.

Carey, S. (1985). *Conceptual change in childhood*. Cambridge, MA: MIT Press.

Chi, M. T., De Leeuw, N., Chiu, M. H., & Lavancher, C. (1994). Eliciting self-explanations improves understanding. *Cognitive Science, 18*(3), 439–477.

Cofré, H., Jiménez, J., Santibáñez, D., & Vergara, C. (2014). Chilean pre-service and in-service teachers and undergraduate students' understandings of evolutionary theory. *Journal of Biological Education*. Advance online publication. doi:10.1080/00219266.2014.967278

DeLoache, J. S. (2004). Becoming symbol-minded. *Trends in Cognitive Sciences, 8*(2), 66–70. doi:10.1016/j.tics.2003.12.004

Department for Education. (2014, July). *National curriculum for England*. Retrieved from https://www.gov.uk/government/collections/national-curriculum

Emmons, N. A., & Kelemen, D. A. (2015). Young children's acceptance of within-species variation: Implications for essentialism and teaching evolution. *Journal of Experimental Child Psychology, 139*, 148–160. doi:10.1016/j.jecp.2015.05.011

Emmons, N. A., Lees, K., & Kelemen, D. A. (2016). *Young children's near and far transfer of the basic theory of natural selection: An analogical storybook intervention*. Manuscript in submission.

Evans, E. M. (2000). The emergence of beliefs about the origins of species in school-age children. *Merrill-Palmer Quarterly, 46*(2), 221–254.

Evans, E. M. (2001). Cognitive and contextual factors in the emergence of diverse belief systems: Creation versus evolution. *Cognitive Psychology, 42*(3), 217–266. doi:10.1006/cogp.2001.0749

Evans, E. M., Legare, C., & Rosengren, K. (2011). Engaging multiple epistemologies: Implications for science education. In M. Ferrari & R. Taylor (Eds.), *Epistemology and science education: Understanding the evolution vs. intelligent design controversy* (pp. 111–139). New York, NY: Routledge.

Evans, E. M., Spiegel, A. N., Gram, W., Frazier, B. N., Tare, M., Thompson, S., & Diamond, J. (2010). A conceptual guide to natural history museum visitors' understanding of evolution. *Journal of Research in Science Teaching, 47*(3), 326–353.

Gelman, S. A., & Rhodes, M. (2012). "Two-thousand years of stasis": How psychological essentialism impedes evolutionary understanding. In K. S. Rosengren, S. K. Brem, E. M. Evans, & G. M. Sinatra (Eds.), *Evolution challenges: Integrating research and practice in teaching and learning about evolution* (pp. 3–21). New York, NY: Oxford University Press.

Gentner, D. (1989). The mechanisms of analogical learning. In S. Vosniadou & A. Ortony (Eds.), *Similarity and analogical reasoning* (pp. 199–241). Cambridge, UK: Cambridge University Press.

Gentner, D., Loewenstein, J., & Thompson, L. (2003). Learning and transfer: A general role for analogical encoding. *Journal of Educational Psychology, 95*(2), 393–408. doi:10.1037/0022-0663.95.2.393

Gopnik, A., & Meltzoff, A. N. (1997). *Words, thoughts, and theories*. Cambridge, MA: MIT Press.

Gregory, T. R. (2009). Understanding natural selection: Essential concepts and common misconceptions. *Evolution: Education and Outreach, 2*(2), 156–175.

Griffith, J. A., & Brem, S. K. (2004). Teaching evolutionary biology: Pressures, stress, and coping. *Journal of Research in Science Teaching, 41*(8), 791–809. doi:10.1002/(ISSN)1098-2736

Guliuzza, R. J. (2014, June). *Brainwashing children to "suppress" design intuition*. Retrieved from http://www.icr.org/article/brainwashing-children-suppress-design/

Herrmann, P. A., French, J. A., DeHart, G. B., & Rosengren, K. S. (2013). Essentialist reasoning and knowledge effects on biological reasoning in young children. *Merrill-Palmer Quarterly, 59*, 198–220. doi:10.1353/mpq.2013.0008

Inagaki, K., & Hatano, G. (2002). *Young children's naive thinking about the biological world.* New York, NY: Psychology Press.

Järnefelt, E., Canfield, C. F., & Kelemen, D. (2015). The divided mind of a disbeliever: Intuitive beliefs about nature as purposefully created among different groups of non-religious adults. *Cognition, 140,* 72–88. doi:10.1016/j.cognition.2015.02.005

Kampourakis, K. (2013). Teaching about adaptation: Why evolutionary history matters. *Science & Education, 22*(2), 173–188. doi:10.1007/s11191-011-9363-2

Keil, F. C. (1989). *Concepts, kinds, and conceptual development.* Cambridge, MA: MIT Press.

Kelemen, D. (1999a). The scope of teleological thinking in preschool children. *Cognition, 70*(3), 241–272. doi:10.1016/S0010-0277(99)00010-4

Kelemen, D. (1999b). Why are rocks pointy? Children's preference for teleological explanations of the natural world. *Developmental Psychology, 35*(6), 1440–1452. doi:10.1037/0012-1649.35.6.1440

Kelemen, D. (2004). Are children "intuitive theists"? Reasoning about purpose and design in nature. *Psychological Science, 15*(5), 295–301. doi:10.1111/psci.2004.15.issue-5

Kelemen, D. (2012). Teleological minds: How natural intuitions about agency and purpose influence learning about evolution. In K. S. Rosengren, S. K. Brem, E. M. Evans, & G. M. Sinatra (Eds.), *Evolution challenges: Integrating research and practice in teaching and learning about evolution* (pp. 66–92). New York, NY: Oxford University Press.

Kelemen, D., Callanan, M. A., Casler, K., & Pérez-Granados, D. R. (2005). Why things happen: Teleological explanation in parent-child conversations. *Developmental Psychology, 41*(1), 251–264. doi:10.1037/0012-1649.41.1.251

Kelemen, D., Emmons, N. A., Seston Schillaci, R., & Ganea, P. A. (2014). Young children can be taught basic natural selection using a picture-storybook intervention. *Psychological Science, 25*(4), 893–902. doi:10.1177/0956797613516009

Kelemen, D., & Rosset, E. (2009). The human function compunction: Teleological explanation in adults. *Cognition, 111*(1), 138–143. doi:10.1016/j.cognition.2009.01.001

Legare, C. H., Lane, J., & Evans, E. M. (2013). Anthropomorphizing science: How does it affect the development of evolutionary concepts? *Merrill-Palmer Quarterly, 59,* 168–197. doi:10.1353/mpq.2013.0009

LoBue, V., Bloom Pickard, M., Sherman, K., Axford, C., & DeLoache, J. S. (2013). Young children's interest in live animals. *British Journal of Developmental Psychology, 31*(1), 57–69. doi:10.1111/j.2044-835X.2012.02078.x

Mayer, R. E., & Moreno, R. (2003). Nine ways to reduce cognitive load in multimedia learning. *Educational Psychologist, 38*(1), 43–52. doi:10.1207/S15326985EP3801_6

Metz, K. E. (1995). Reassessment of developmental constraints on children's science instruction. *Review of Educational Research, 65*(2), 93–127. doi:10.3102/00346543065002093

Moore, C., & Dunham, P. J. (1995). *Joint attention: Its origins and role in development.* Hillsdale, NJ: Erlbaum.

National Research Council. (2012). *A framework for K-12 science education: Practices, crosscutting concepts, and core ideas.* Washington, DC: National Academies Press.

Nehm, R. H., Kim, S. Y., & Sheppard, K. (2009). Academic preparation in biology and advocacy for teaching evolution: Biology versus non-biology teachers. *Science Education, 93*(6), 1122–1146. doi:10.1002/sce.v93:6

Nehm, R. H., & Schonfeld, I. S. (2007). Does increasing biology teacher knowledge of evolution and the nature of science lead to greater preference for the teaching of evolution in schools? *Journal of Science Teacher Education, 18* (5), 699–723. doi:10.1007/s10972-007-9062-7

Poling, D. A., & Evans, E. M. (2002). Why do birds of a feather flock together? Developmental change in the use of multiple explanations: Intention, teleology and essentialism. *British Journal of Developmental Psychology, 20*(1), 89–112. doi:10.1348/026151002166343

Poling, D. A., & Evans, E. M. (2004). Are dinosaurs the rule or the exception? Developing concepts of death and extinction. *Cognitive Development, 19*(3), 363–383. doi:10.1016/j.cogdev.2004.04.001

Rosengren, K. S., Brem, S. K., Evans, E. M., & Sinatra, G. M. (Eds.). (2012). *Evolution challenges: Integrating research and practice in teaching and learning about evolution.* Oxford, UK: Oxford University Press.

Rosengren, K. S., Gelman, S. A., Kalish, C. W., & McCormick, M. (1991). As time goes by: Children's early understanding of growth in animals. *Child Development, 62,* 1302–1320. doi:10.2307/1130808

Rutledge, M. L., & Warden, M. A. (2000). Evolutionary theory, the nature of science and high school biology teachers: Critical relationships. *The American Biology Teacher, 62*(1), 23–31. doi:10.1662/0002-7685(2000)062[0023:ETTNOS]2.0.CO;2

Samarapungavan, A., & Wiers, R. W. (1997). Children's thoughts on the origin of species: A study of explanatory coherence. *Cognitive Science, 21*(2), 147–177. doi:10.1207/s15516709cog2102_2

Schweingruber, H. A., Duschl, R. A., & Shouse, A. W. (Eds.). (2007). *Taking science to school: Learning and teaching science in grades K-8.* Washington, DC: National Academies Press.

Shtulman, A., & Schulz, L. (2008). The relation between essentialist beliefs and evolutionary reasoning. *Cognitive Science, 32,* 1049–1062. doi:10.1080/03640210801897864

Siegal, M., & Peterson, C. (1999). *Children's understanding of biology and health.* New York, NY: Cambridge University Press.

Sinatra, G. M., Brem, S. K., & Evans, E. M. (2008). Changing minds? Implications of conceptual change for teaching and learning about biological evolution. *Evolution: Education and Outreach, 1*(2), 189–195.

Spiegel, A., Evans, E. M., Frazier, B. F., Hazel, A., Tare, M., Gram, W., & Diamond, J. (2012). Changing museum visitors' concepts of evolution. *Evolution: Education and Outreach*, *5*, 43–61.

Tare, M., Chiong, C., Ganea, P., & DeLoache, J. (2010). Less is more: How manipulative features affect children's learning from picture books. *Journal of Applied Developmental Psychology*, *31*(5), 395–400. doi:10.1016/j.appdev.2010.06.005

Tomasello, M., & Farrar, M. J. (1986). Joint attention and early language. *Child Development*, *57*(6), 1454–1463. doi:10.2307/1130423

Wellman, H. M., & Gelman, S. A. (1992). Cognitive development: Foundational theories of core domains. *Annual Review of Psychology*, *43*(1), 337–375. doi:10.1146/annurev.ps.43.020192.002005

Zohar, A., & Ginossar, S. (1998). Lifting the taboo regarding teleology and anthropomorphism in biology education? Heretical suggestions. *Science Education*, *82*(6), 679–697. doi:10.1002/(ISSN)1098-237X

Children's Ability to Learn Evolutionary Explanations for Biological Adaptation

Andrew Shtulman, Cara Neal, and Gabrielle Lindquist

ABSTRACT

Research Findings: Evolution by natural selection is often relegated to the high school curriculum on the assumption that younger students cannot grasp its complexity. We sought to test that assumption by teaching children ages 4–12 (n = 96) a selection-based explanation for biological adaptation and comparing their success to that of adults (n = 30). Participants provided explanations before and after a 10-min, analogy-based tutorial illustrating the principles of variation, differential survival, differential reproduction, inheritance, and population change. Although younger children (ages 4–6) showed minimal evidence of learning these principles, older children (ages 7–12) showed robust evidence of doing so, learning them at rates equivalent to adults. Participants of all ages, however, provided nonevolutionary explanations for biological adaptations (i.e., explanations referencing need, growth, and creation) nearly as often at posttest as they did at pretest. *Practice or Policy*: These results suggest that older elementary school-age children can be taught evolutionary concepts but that learning such concepts does not lead to the automatic replacement of nonevolutionary views of biological adaptation, which must be addressed separately.

Evolution by natural selection is the theory that unifies all biological phenomena, yet evolution is rarely taught to children. Children are expected to begin their biology education without reference to the very principles that structure and organize the field. The U.S. National Research Council, for instance, recommends that elementary school students learn about anatomy, physiology, taxonomy, and ecology but not evolution (National Academy of Sciences, 2013). Evolution is recommended for inclusion only in the high school curriculum. Ideas *related to* evolution are recommended for inclusion in the elementary school curriculum (e.g., heredity, biodiversity, trait variation, extinction), but evolution itself is waylaid for several years. The consequence of this recommendation is that children learn about biological systems (e.g., life cycles, food chains, symbiosis) and biological processes (e.g., digestion, respiration, reproduction) without reference to the historical pressures that gave rise to those phenomena and thus without explanation for their form or function.

Some children may never learn those explanations, even in high school, as evolution is either taught poorly or not taught at all. A recent survey of U.S. high school biology teachers found that only 28% teach evolution as an uncontroversial fact, supported by ample data. Most teachers (60%) avoid the topic as much as possible, and some (12%) explicitly advocate for creationism (Berkman & Plutzer, 2011; see also Nehm, Kim, & Sheppard, 2010). The scientific community has recognized, for several decades now, that "nothing in biology makes sense except in the light of evolution" (Dobzhansky, 1973), yet many students learn biology in the absence of such light.

One reason why children are shielded from the concept of evolution is that evolution is a controversial topic (Blancke, De Smedt, De Cruz, Boudry, & Braeckman, 2012; Lombrozo,

Shtulman, & Weisberg, 2006). Most U.S. citizens deny that humans have evolved from nonhuman ancestors strictly by natural means, favoring either creationist explanations or quasi-creationist explanations in which God guides evolutionary change (Newport, 2010). Many people also see evolution as antithetical to positive social values, like altruism, equality, and self-determination (Brem, Ranney, & Schindel, 2003). But controversy surrounding evolution is not the only reason why children are shielded from the concept. Another is that evolution is difficult to understand, and children are viewed by many as cognitively ill-equipped to handle that complexity (Carmichael, 2009).

Dozens of studies have shown that evolution is difficult to understand even for college-educated adults (for reviews, see Gregory, 2009; Kampourakis, 2014; Shtulman & Calabi, 2012). Evolution, or population-level changes in the frequency of heritable traits, is driven by selection, or the differential survival and reproduction of the varied organisms within a population, yet most adults do not conceive of evolution in these terms. Rather, most adults conceive of evolution as the uniform transformation of all members of the species, with each organism producing offspring more adapted to the environment than the organism itself was at birth. According to this view, mutations are not random; they occur in a direction that would benefit the organism, and all organisms within the population are assumed to acquire the same mutations at the same time (Bishop & Anderson, 1990; Brumby, 1984; Shtulman, 2006; Shtulman & Calabi, 2013; Shtulman & Schulz, 2008).

Underlying this misconceived view of evolution are three widespread and early developing inductive biases: teleology, essentialism, and intentionality. *Teleology* is the idea that an organism's traits can be explained by their function (Kelemen, 2012; Lombrozo & Carey, 2006). Teleology allows people to recognize that wings exist for flying and that lungs exist for breathing, but it causes them to ignore the historical origin of those traits (i.e., selection over a population of individuals who once possessed only the precursors of those traits). *Essentialism* is the idea that an organism's traits are byproducts of its hidden nature or essence (Gelman, 2003; Gelman & Rhodes, 2012). Essentialism allows people to recognize that gray baby flamingos will grow to be pink and that flat-nosed baby rhinos will grow to have horns, but it causes them to ignore variation in those traits and the consequences of such variation (i.e., differential survival and differential reproduction). And *intentionality* is the idea that organisms act on the environment in ways that further their goals and desires (Evans, 2008; Johnson, 2000). Intentionality allows people to recognize that holes in the beach are the handiwork of crabs and dams in the river are the handiwork of beavers, but it causes them to interpret crabs and beavers as intentionally designed themselves (e.g., by a divine force or divine intelligence). Intentionality can also cause people to assume that evolution is driven by the intentions of the evolving organisms (Legare, Lane, & Evans, 2013).

Each of these biases serves a valuable function when applied to the properties of individual organisms but can lead to misconceptions when applied to the properties of entire species, causing adults to misconceive evolution (Sinatra, Brem, & Evans, 2008). These biases are widespread and early developing, emerging prior to any formal instruction in biology (Coley & Tanner, 2012). Children are thus vulnerable to the same kinds of evolutionary misconceptions as those observed among adults. Indeed, young elementary school-age children have been shown to prefer intentional (i.e., creationist) explanations of biological adaptation to evolutionary explanations (Evans, 2001), and older elementary school-age children have been shown to prefer essentialist explanations (Samarapungavan & Wiers, 1997) and teleological explanations (Berti, Toneatti, & Rosati, 2010) to strictly selection-based explanations.

Evolution is thus difficult to grasp at all ages, but no study has shown that evolution is *more difficult* to grasp at younger ages than at older ages. Evolution may actually be easier to grasp at younger ages, before children have acquired as much experience viewing the biological world through the lenses of teleology, essentialism, and intentionality. When students are introduced to evolution in high school (or college), they have spent more than a decade of their lives reasoning about biological phenomena in explicitly nonevolutionary terms. Introducing younger students to evolution could potentially bypass this obstacle, as they have yet to encode as much biological information in those terms. Children are predisposed to construing biological traits as purpose-based, nonvariable, and intentionally designed, just as adolescents and adults are, but they may have less conceptual baggage to work through on the whole (see Kelemen, Emmons, Schillaci, & Ganea, 2014, for a similar argument).

To date, no studies have compared children's receptiveness to evolutionary ideas to that of adults. However, two studies have explored whether children can be taught evolutionary ideas at all. In the first, Legare et al. (2013) provided children between the ages of 5 and 12 with explanations for biological adaptation that illustrated five key evolutionary principles: variation, inheritance, selection, time, and adaptation. Some children received explanations cloaked in anthropomorphic, desire-based language (e.g., "some eagles *wanted* to change," "they *tried* to be bigger"); some received explanations cloaked in teleological, need-based language (e.g., "some eagles *needed* to change," "they *needed* to be bigger"); and some received explanations containing only selection-based language (e.g., "chicks that were bigger lived longer, and they had more chicks").

After listening to three sample explanations, children were asked to reiterate those explanations. Children who had received desire-based explanations recalled significantly fewer evolutionary principles than those who had received need-based or selection-based explanations. These findings indicate that children can encode and recall evolutionary explanations, particularly if those explanations are framed in non-anthropomorphic terms, but they do not indicate whether children can generate those explanations on their own.

Following Legare et al. (2013), Kelemen et al. (2014) documented that children can indeed generate evolutionary explanations on their own. These researchers read a storybook illustrating an evolutionary explanation of biological adaptation to children between the ages of 5 and 8. The biological adaptation was novel—the evolution of thin trunks in a fictional species of elephant-like creatures ("pilosas")—and the explanation for its origin entailed six concepts: trait variation, resource limitation, differential survival, differential reproduction, trait inheritance, and population-level changes in trait frequency. Following the storybook, children were asked to explain two new instances of biological adaptation, one presented that same day (an immediate posttest) and another presented 3 months later (a delayed posttest).

Children's explanations were elicited in the form of a clinical interview consisting of five closed-ended questions and five open-ended questions. Responses to the interview were coded for one of five levels of understanding, ranging from no retention of the instructional materials (Level 0), to retention of isolated facts (Level 1), to understanding differential survival but not differential reproduction (Level 2), to understanding differential survival and differential reproduction in one generation but not multiple generations (Level 3), to understanding differential survival and differential reproduction across multiple generations and thus how these processes give rise to population-level change (Level 4).

The researchers found that children's understanding of evolution increased dramatically with the intervention. Before the intervention, only around 10% of 5- and 6-year-olds and 40% of 7- and 8-year-olds were able to articulate a Level 2 understanding of evolution or better. After the intervention, those figures rose to 40% and 90%, respectively, both at the immediate posttest and delayed posttest. That said, few children were able to articulate a Level 4 understanding at posttest; most went from a Level 0 understanding to a Level 2 understanding.

Kelemen et al.'s (2014) findings are promising with respect to the pedagogical goal of introducing evolution earlier in school, but those findings are limited in several respects. First, Kelemen and colleagues (2014) did not compare children's ability to learn evolutionary principles to that of adults, and it remains unclear whether children are more or less adept at this. Second, Kelemen and colleagues (2014) assessed children's understanding of the evolution of only one type of trait—traits that aid foraging—and it remains unclear whether children would be able to apply that understanding to other types of traits, like traits that aid predator avoidance. Third, Kelemen and colleagues (2014) coded children's explanations as a single unit, and it remains unclear whether children were capable of learning all evolutionary concepts equally well or were inclined to learn some concepts (e.g., differential survival) better than others (e.g., within-species variation).

Fourth, and most important, Kelemen and colleagues (2014) probed children's understanding of evolution with up to 10 different prompts, and it remains unclear whether children would have been equally successful with fewer prompts. For instance, children who mentioned trait variation but not differential survival were prompted for more information along those lines (e.g., "Which group of pilosas got more food?"). Likewise, children who mentioned differential survival but not differential reproduction were prompted for more information (e.g., "Which group of pilosas had more babies?"), as were children who mentioned differential survival and reproduction in one generation but not differential survival and reproduction across several generations (e.g., "If this pilosa had a baby, what kind of trunk would its baby have?").

In the present study, we sought to replicate Kelemen et al.'s (2014) findings, assessing children's explanations for biological adaptation before and after a storybook-based tutorial on evolution by natural selection, but we also sought to address the aforementioned limitations in their study. We did so by implementing the following changes. First, we explicitly compared children's ability to learn evolutionary principles to that of adults, controlling for differences in prior knowledge at pretest. Second, we varied the type of traits under consideration, including not only foraging-related traits (which increase an organism's chances of finding food) but also camouflage-related traits (which decrease an organism's chances of *becoming* food). Third, we coded participants' explanations analytically rather than holistically, tabulating how frequently they mentioned each of the five evolutionary principles illustrated in the tutorial. Fourth, we elicited children's explanations for biological adaptation with a single prompt, namely, "How did [this particular animal] come to have [this particular trait]?"

One additional change we made to Kelemen et al.'s (2014) methodology was to include a second training example in our tutorial, presenting children with two illustrations of evolution rather than one. We added a second example for several reasons. First, we wanted to vary whether our training examples covered the same type of trait (two foraging-related traits) or different types of traits (one foraging-related trait and one camouflage-related trait), which we discuss in more detail later. Second, we wanted to create an opportunity to measure how well children abstracted a general schema from the specific examples at hand, and we did so by asking them whether they noticed any commonalities between the two training examples and, if so, what. Including a measure of schema abstraction is standard practice in the analogy literature (Gentner, Loewenstein, & Thompson, 2003; Novick & Holyoak, 1991), as it can provide direct evidence of the relation between schema abstraction and schema transfer.

Finally, we were concerned that a single training example would be ineffective at illustrating an abstract schema. Although a single example proved sufficient in Kelemen et al.'s (2014) study, other studies have shown that multiple examples are typically more effective than a single example, particularly if participants are encouraged to compare those examples (Catrambone & Holyoak, 1989; Christie & Gentner, 2010; Kurtz, Miao, & Gentner, 2001; Son, Smith, & Goldstone, 2011). By comparing examples, learners are able to ignore superficial differences in the examples' content and focus on the more substantive commonalities in their underlying structure. This form of instruction, known as *analogical encoding*, has proven effective at helping learners abstract "a coherent and portable relational structure" from specific instantiations of that structure (Gentner et al., 2003, p. 402).

Given these changes to Kelemen et al.'s (2014) methodology, we predicted that children would still be able to learn evolutionary principles but were unclear on whether children would be able to learn those principles as well as adults. We were also unclear on whether children would be able to learn all evolutionary principles equally well and whether they would be able to apply those principles to all traits equally well. Nevertheless, we strongly expected that children's ability to abstract an evolutionary schema from the training examples would predict their ability to apply that schema to new instances of biological adaptation, not only because previous studies have demonstrated similar effects in other domains (e.g., Gentner et al., 2003; Novick & Holyoak, 1991) but also because our decision to elicit evolutionary explanations with minimal prompting put the onus of generating such explanations squarely on the participant.

Method

Participants

The participants were 96 children recruited from playgrounds in Pasadena, California, and 30 college undergraduates recruited from introductory psychology courses at Occidental College. The children ranged in age from 4 to 12 and were sampled from a racially diverse population (42% Caucasian, 25% Latino, 10% African American, 10% Asian, and 13% other). For the purpose of data analysis, we split the children into two groups: those between the ages of 4 and 6 (*M* age = 5 years, 6 months; range = 4 years, 0 months, to 6 years, 9 months) and those between the ages of 7 and 12 (*M* age = 9 years, 0 months; range = 7 years, 0 months, to 12 years, 8 months).

Children in the first age group (n = 52) are referred to as *younger children*, and those in the second (n = 44) are referred to as *older children*. The age range for the older children was twice as large as that for the younger children only because we did not want to exclude the 10% of children who participated in our study but happened to be older than the age of 9. (Recruiting participants at parks by sight did not allow us to be as restrictive with age as is possible with registry-based or school-based methods of recruitment.) Excluding those children from the analysis would not, however, have altered any of the reported differences between our younger group and our older group.

Approximately half of the children in each group were girls and half were boys. Preliminary analyses revealed no effects of gender on either pretest scores or posttest scores, so gender was not included as a variable in the final analyses.

With respect to the adults, they ranged in age from 18 to 22 and came from a variety of academic majors, though primarily social science majors. Prior to the study, they had taken an average of 1.6 biology courses in high school and 1.4 biology courses in college. All had taken at least one high school–level course, and 14 of the 30 had taken at least one college-level course. In addition, 21 were women and nine were men. They were compensated with extra credit in a psychology or cognitive science course.

Procedure

All participants—children and adults alike—received the assessment and training materials in the form of a picture book. The first section of picture book contained two pretest items, the second contained two training items, and the third contained two posttest items. Pretest and posttest items consisted of a picture of an animal, a description of one of its traits, and a prompt to explain where that trait came from. An example is as follows:

> This is a panda. Pandas have thumbs. Why do you think they have thumbs? Pandas have thumbs so that they can grab bamboo, which is the only food they eat. Did you know that the ancestors of pandas—who lived long, long ago—did not have thumbs? How do you think they came to have thumbs?

We provided children with function information about each trait before asking them to explain the origin of that trait because previous studies have shown that children treat only functional traits as heritable (Springer & Keil, 1989; Ware & Gelman, 2014). That said, providing children with function information may have primed teleological conceptions of trait inheritance, thereby limiting children's access to evolutionary ones. Emmons and Kelemen (2015), for instance, have found that children are less likely to accept that a trait can vary across members of the same species if that trait is described in functional terms. Future studies are needed to determine whether the inclusion of function information did more harm than good, as this type of information could easily be excluded.

Participants were asked to explain the origin of one foraging-related trait and one camouflage-related trait at each assessment period (pretest and posttest). The foraging-related traits were the thumb of the panda and the beak of the pelican, and the camouflage-related traits were the striped coat of the zebra and the leaf-like wing of the katydid. Each trait was included as a pretest item for half of the children and as a posttest item for the other half, and the ordering of those traits was counterbalanced across participants.

The training items, like the assessment items, focused participants' attention on a particular animal and a particular trait, but they differed from the training items in that participants were not asked to explain the origin of the trait; they were provided with an explanation by the experimenter. Each explanation entailed the combination of five evolutionary principles: variation, differential survival, differential reproduction, inheritance, and population change. The following is a sample training item, with each principle listed in brackets following its illustration:

> This is a Guinea turaco. Guinea turacos are green. Why do you think they're green? Guinea turacos are green so they can blend into the jungle and avoid being eaten by hawks and snakes. The ancestors of Guinea turaco—who lived long, long ago—were not green. They were gray. Let me tell you how green turacos came from gray turacos. Once, by chance, some turacos were born with green feathers [*variation*]. The green turacos lived longer than the gray turacos because they were better able to hide from hawks and snakes [*differential survival*]. Because the green turacos lived longer than the gray turacos, the green turacos had more babies [*differential reproduction*]. The babies of the green turacos had green feathers, just like their parents [*inheritance*]. After many, many years, all the gray turacos were replaced by green ones [*population change*].

Each evolutionary principle included in the training items was illustrated visually. Variation was illustrated with a picture of 10 turacos in which nine were gray and one was green. Differential survival was illustrated with a picture of two arrows depicting relative lifespan: a long arrow appearing beneath a green turaco and a short arrow appearing beneath a gray turaco. Differential reproduction was illustrated with a picture of two broods of baby turacos: a small brood of gray turacos and a large brood of green turacos. Inheritance was illustrated with a picture of the parent turaco for each brood. And population change was illustrated with a series of five pictures. Each picture displayed 10 turacos, in which the ratio of green turacos to gray turacos changed from 1:9 to 3:7 to 5:5 to 8:2 to 10:0.

Children were presented with two training items back to back. The training items covered either foraging-related traits (the long snout of the anteater, the long neck of the giraffe) or camouflage-related traits (the green feathers of a Guinea turaco, the white fur of the Arctic hare). A third of the children were shown two foraging-related traits, a third were shown two camouflage-related traits, and a third were shown one foraging-related trait and one camouflage-related trait.

We had predicted that children who received a mixed pair of traits would learn more than those who received only foraging-related traits or only camouflage-related traits, as other studies have shown that materials with dissimilar surface-level content are more pedagogically effective than materials with similar surface-level content (e.g., Gick & Holyoak, 1983; Vasilyeva & Bowers, 2010). However, we found no such effect either on posttest scores as a whole or on posttest scores for particular types of traits. We therefore pooled the data across training conditions.

At the end of the training phase, participants were asked whether they had noticed any similarities between the two training items. Participants' responses were coded for evidence of the five illustrated principles—variation, differential survival, differential reproduction, inheritance, and population change—and constituted a measure of schema abstraction. Note that schema abstraction was not our focal measure of learning. Our focal measure was how many more evolutionary principles participants cited at posttest compared to pretest. This measure captures participants' ability to apply the five-principle evolutionary schema, conveyed during training, to new examples of biological adaptation after controlling for baseline differences in prior knowledge of that schema.

Regardless of whether participants were able to abstract the schema on their own, the experimenter articulated that schema before moving to the posttest. For instance, in the training condition involving two camouflage-related traits, the experimenter told participants the following:

> Did you notice any similarities between the story about the turacos and the story about the hares? Did both kinds of animals change over time for similar reasons? They did! In both stories, there were differences among the animals, and the animals that could hide better from predators lived longer than the other animals. And because they lived longer, they had more babies. And those babies had the same traits as their parents. After many, many years, the animals that could hide better from predators replaced the other animals.

The tutorial took approximately 10 min, and the entire interview took approximately 20 min. Across the interview, participants provided a total of five explanations for biological adaptation: two at pretest, one at training, and two at posttest. All explanations were audio recorded and transcribed at a later date.

It should be noted that the same protocol was used for both children and adults, meaning that adults, like children, provided their explanations verbally in the context of a one-on-one interview with the experimenter. We collected adults' responses in this manner to minimize differences in the format of our training and assessment materials across age groups and, accordingly, format-related differences in responding.

Coding

Participants' explanations were coded for inclusion of the five evolutionary principles taught during the training period. The following explanation, provided by an 11-year-old girl, is an example of an explanation that included all five:

> One of them probably had thumbs [*variation*], and they were able to eat more bamboo, and so they lived longer [*differential survival*], and they passed on the genes because they lived longer [*inheritance*]. Their babies, since they had more babies than the other ones [*differential reproduction*], lived longer too because they had thumbs and they had more babies and more babies until all of them had thumbs. All the pandas without thumbs died because they couldn't get enough food [*population change*].

Participants' explanations were also coded for alternative, nonevolutionary considerations. The three most commonly cited considerations were need, growth, and creation. *Need-based explanations* appealed to the needs of an organism as an explanation, in and of itself, for the origin of a trait that met those needs. Examples include "because they need to hold the bamboo" and "because they need to catch fish." *Growth-based explanations* appealed to ontogenetic development as an explanation, in and of itself, for the origin of a new trait. Examples include "they got older and grew them" and "they just grew thumbs; they grew fur and then they grew thumbs." Finally, *creation-based explanations* appealed to a divine creator as the source of the trait. Examples include "because God gave them thumbs" and "because God made them like that." In total, 52% of children and 60% of adults provided at least one need-based explanation, 43% of children and 17% of adults provided at least one growth-based explanation, and 19% of children (but none of the adults) provided at least one creation-based explanation.

Our reasons for focusing on need-based, growth-based, and creation-based explanations were not only empirical but also theoretical. Each type of explanation exemplifies a pattern of reasoning discussed earlier as an impediment to evolutionary reasoning. Need-based explanations exemplify *teleological* reasoning in that they cite a trait's purpose as an explanation for its existence. Growth-based explanations exemplify *essentialist* reasoning in that they cite an organism's innate potential to develop a certain trait as an explanation for the origin of that trait. And creation-based explanations exemplify *intentional* reasoning in that they cite the intentions of a divine creator as an explanation for the creator's handiwork.

One final piece of data extracted from the interview transcripts was how often the experimenters prompted participants for additional information. These prompts took two forms: general prompts and schema reminders. *General prompts* included requests for an answer ("Can you make a guess?"), requests for an elaboration ("Can you tell me more?"), and requests for additional ideas in cases in which participants spontaneously cited only nonevolutionary considerations ("Can you think of any other way pandas came to have thumbs?"). *Schema reminders*, in contrast, were explicit requests to relate the posttest items to the training items and the principles illustrated therein ("Can you think of a way zebras got their stripes that's similar to how Guinea turacos got their green feathers and Arctic hares got their white fur?"). Both types of prompts were tabulated and compared to participants' responses to determine whether they influenced how often participants cited evolutionary considerations at each assessment period.

All explanations were coded by two independent coders. Intercoder agreement was uniformly high across the 10 coding categories: 91% agreement for how often participants mentioned variation, 90% for differential survival, 95% for differential reproduction, 97% for inheritance, 94% for population change, 87% for need, 96% for growth, 98% for creation, 81% for general prompts, and 98% for schema reminders. All disagreements were resolved through discussion.

Results

Here we analyze participants' evolutionary explanations for effects of age and training, followed by their nonevolutionary explanations. Composite scores for the two analyses had different ranges given that we coded for different numbers of evolutionary and nonevolutionary considerations. Composite scores for the first analysis could range from 0 to 10 at each assessment period (five principles across two items), whereas those for the second analysis could range from 0 to 6 (three considerations across two items).

Effects of training on evolutionary explanations

The mean number of evolutionary principles cited before and after training is displayed in Figure 1 as a function of age. Participants in all age groups benefited from training, though younger children benefited less than did older children and adults. The reliability of these effects was assessed with a repeated measures analysis of variance in which assessment period (pretest vs. posttest) was treated as a within-participants factor and age group (younger children vs. older children vs. adults) was treated as a between-participants factor. This analysis revealed a significant effect of assessment period, a significant effect of age group, and a significant interaction between assessment period and age group: assessment period, $F(1, 123) = 95.02$, $p < .001$; age group, $F(2, 123) = 14.79$, $p < .001$; assessment period × age group, $F(2, 123) = 47.35$, $p < .001$.

We explored the main effect of age group with Bonferroni comparisons, which revealed that younger children cited significantly fewer evolutionary principles than did older children and that older children cited significantly fewer evolutionary principles than did adults. We explored the interaction with paired-samples *t* tests comparing pretest scores to posttest scores for each age group by itself. These analyses revealed significant effects for all three groups: younger children, $t(51) = 2.91$, $p < .01$; older children, $t(43) = 6.23$, $p < .001$; adults, $t(29) = 5.94$,

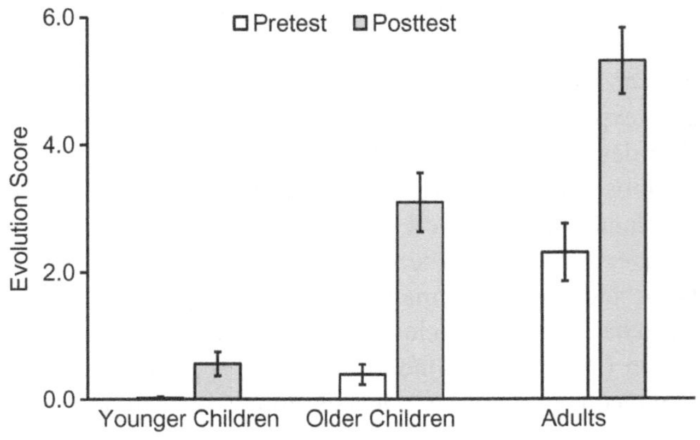

Figure 1. Mean evolution scores as a function of age and training (range = 0–10). Error bars represent the standard error.

$p < .001$. Nevertheless, the size of this effect was substantially smaller for younger children ($d = 0.55$) than it was for either older children ($d = 1.19$) or adults ($d = 1.13$).

This last set of analyses demonstrates that participants cited more evolutionary considerations after training than before, but did training affect their ability to cite all considerations equally? We addressed this question by comparing pretest scores to posttest scores for each evolutionary principle on its own. The results of this analysis are displayed in Table 1. Younger children were more likely to cite three of the five principles at posttest than at pretest (variation, differential survival, and differential reproduction); older children and adults were significantly more likely to cite all five. Thus, the effects of training appeared to be widespread.

The data in Table 1 suggest not only that effects of training were widespread but also that they were similar for older children and adults. To verify this observation statistically, we submitted participants' posttest scores to an analysis of covariance in which age group (younger children vs. older children vs. adults) was treated as a between-subjects factor and pretest score was treated as a covariate. It is not surprising that this analysis revealed a significant effect of age group, $F(2, 122) = 97.52$, $p < .001$. However, follow-up comparisons with Bonferroni corrections revealed that when we controlled for pretest scores, younger children scored significantly lower than did older children (M difference = 2.3, $p < .001$) and adults (M difference = 3.3, $p < .001$), but older children scored the same as adults (M difference = 1.0, *ns*).

Thus, even though adults cited more evolutionary principles than did older children at pretest, the *difference* in the number of principles cited from pretest to posttest was the nearly same. This finding cannot be attributed to a ceiling effect on behalf of the adults because adults' average score at posttest (5.3) was only half of what it could have been (10).

Effects of training on alternative conceptions

The mean number of alternative (nonevolutionary) conceptions cited before and after training is displayed in Figure 2 as a function of age. Across age groups, training had a minimal effect on how often participants cited such considerations. Indeed, a repeated measures analysis of variance revealed no effect of assessment period on the number of alternative conceptions cited, $F(1, 123) = 3.27$, *ns*. It did, however, reveal a significant effect of age group and a near-significant interaction between assessment period and age group: age group, $F(2, 123) = 3.15$, $p < .05$; assessment period × age group, $F(2, 123) = 2.92$, $p < .06$. We explored the main effect of age group with Bonferroni comparisons, which revealed that younger children provided significantly fewer alternative conceptions than did adults but that older children provided approximately the same number

Table 1. The Mean Frequency With Which Each Evolutionary Principle Was Mentioned at Each Assessment Period by Participants in Each Age Group (Range = 0–2).

Group	*Concept*	*Pretest*	*Posttest*	*Difference*
Younger children	Variation	0.02	0.13	0.12*
	Differential survival	0.00	0.21	0.21*
	Differential reproduction	0.00	0.12	0.12*
	Inheritance	0.00	0.04	0.04
	Population change	0.00	0.06	0.06
Older children	Variation	0.16	0.84	0.68***
	Differential survival	0.05	0.75	0.71***
	Differential reproduction	0.00	0.41	0.41**
	Inheritance	0.05	0.23	0.18*
	Population change	0.14	0.86	0.72***
Adults	Variation	0.23	1.17	0.93***
	Differential survival	1.13	1.53	0.40*
	Differential reproduction	0.43	1.03	0.60***
	Inheritance	0.30	0.60	0.30*
	Population change	0.20	0.97	0.77***

*$p < .05$. **$p < .01$. ***$p < .001$.

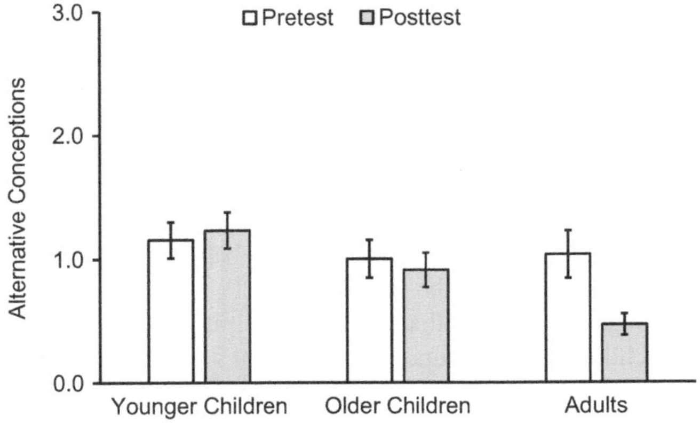

Figure 2. Mean alternative conception scores as a function of age and training (range = 0–6). Error bars represent the standard error.

of alternative conceptions as did both younger children and adults. We explored the interaction effect with paired-samples *t* tests comparing pretest scores to posttest scores for each age group by itself. These analyses revealed a significant pre/post difference for adults but no significant differences for younger children or for older children: adults, $t(29) = 2.81$, $p < .01$; younger children, $t(51) < 1.0$, *ns;* older children, $t(43) < 1.0$, *ns.*

These patterns were investigated at the level of individual conceptions as well. Table 2 displays the frequency with which participants in each age group cited need, growth, and creation at each assessment period. These frequencies remained constant for younger children and older children but decreased for adults in the category of need. Still, the size of this solitary effect was small ($d = 0.53$).

Effects of schema abstraction and schema reminders

During the training phase of the study, participants were asked whether they noticed any similarities between the two training examples before those similarities were highlighted for them. Their responses were coded for the same five evolutionary principles coded for at each assessment period. On average, younger children spontaneously abstracted 0.2 evolutionary principles from the training materials ($SD = 0.4$), older children spontaneously abstracted 1.0 ($SD = 1.1$), and adults spontaneously abstracted 1.5 ($SD = 1.2$). These differences were statistically significant, $F(2, 123) = 18.56$, $p < .001$. Follow-up comparisons with Bonferroni corrections revealed that although younger children abstracted significantly fewer principles than older children, older children did not abstract significantly fewer principles than adults.

Table 2. The Mean Frequency With Which Each Alternative Conception Was Mentioned at Each Assessment Period by Participants in Each Age Group (Range = 0–2).

Group	*Concept*	*Pretest*	*Posttest*	*Difference*
Younger children	Need	0.50	0.65	0.15
	Growth	0.37	0.44	0.08
	Creation	0.29	0.13	−0.15
Older children	Need	0.41	0.32	−0.09
	Growth	0.43	0.52	0.09
	Creation	0.16	0.07	−0.09
Adults	Need	0.87	0.43	−0.43*
	Growth	0.17	0.03	−0.13
	Creation	0.00	0.00	0.00

*$p < .05$.

Individual participants varied widely in how many principles they abstracted, and we sought to determine whether this variation tracked participants' ability to apply those principles at posttest. We did so with a hierarchical regression in which we regressed participants' evolution scores at posttest against their evolution scores at training after first controlling for their evolution scores at pretest. This analysis revealed significant effects of both pretest scores and training scores: pretest scores, $\beta = 0.80$, $t(125) = 5.63$, $p < .001$; training scores, $\beta = 0.96$, $t(125) = 4.32$, $p < .001$. Moreover, training scores explained significantly more variance in posttest scores than did pretest scores alone (R^2 for pretest scores = .28, R^2 for pretest scores + training scores = .37), $\Delta F(1, 123) = 18.68$, $p < .001$. This finding remained largely unchanged even when adults were excluded from the analysis (R^2 for pretest scores = .14, R^2 for pretest scores + training scores = .23), $\Delta F(1, 93) = 11.04$, $p < .001$. In short, the more evolutionary principles participants spontaneously abstracted at training, the more evolutionary principles they cited at posttest, regardless of how many evolutionary principles they cited at pretest.

At posttest, if participants failed to provide an evolutionary explanation, they were encouraged to think back to the training examples and the principles illustrated by those examples. Younger children required the same number of schema reminders as older children (younger children: $M = 1.2$, $SD = 0.1$; older children: $M = 1.2$, $SD = 0.1$), and adults never required such reminders, as they cited at least some evolutionary principles on each posttest item. Schema reminders did not, however, prove helpful. The more schema reminders children required, the fewer evolutionary principles they cited, $r(94) = -.30$, $p < .01$. In other words, children who provided evolutionary explanations at posttest did so without prompting, and children who provided nonevolutionary explanations at posttest did not provide evolutionary ones following prompting.

General prompts (e.g., "Can you make a guess?" "Can you tell me more?") were also ineffective at eliciting evolutionary explanations, as their frequency was not correlated with children's evolution scores either at pretest or at posttest: pretest, $r(94) = .05$, *ns;* posttest, $r(94) = .05$, *ns*. The prompts in our study were thus minimal in both number and effect.

Discussion

Can elementary school-age children learn evolutionary explanations for biological adaptation? The answer appears to be yes, at least for children on the older side of this age range. When left to their own devices, children tend to explain biological adaptation in terms of need, growth, or creation. However, after being shown two instances of how differential survival and differential reproduction of the varied organisms within a population can lead to heritable changes across the entire population, children were able to articulate those principles in general terms and apply them to new instances of adaptation, particularly children over the age of 7. In fact, children of this age were able to apply newly learned evolutionary principles as well as adults. Adults may have cited more evolutionary principles at pretest, but both groups increased their propensity to cite evolutionary principles, from pretest to posttest, by about the same amount.

Our findings replicate Kelemen et al.'s (2014) findings that elementary school-age children can learn evolutionary principles from a brief, storybook-based tutorial. And they extend those findings by showing that children's receptivity to evolutionary information encompasses different types of traits (not just foraging-related traits) and different types of principles (when knowledge of those principles is assessed separately, principle by principle). They also extend those findings by showing that children's receptivity to evolutionary information tracks their ability to abstract an evolutionary schema from specific instances of evolutionary change, thus pointing to analogical encoding as a possible method for teaching evolutionary concepts.

However, our findings do not replicate one key finding from Kelemen et al.'s (2014) study: that children younger than the age of 6 are (robustly) capable of learning evolutionary explanations. In Kelemen et al.'s (2014) study (Study 1), children younger than the age of 6 increased their scores on an evolution assessment by an average of 1.1 points on a 4-point scale (or 28%), whereas children of the same age in our study increased their scores by only 0.6 points on a 5-point scale (or 12%). The difference was reliable but too small to be of much practical significance. Only 15% of our younger children increased their score by 1 point or more.

What factors might account for this discrepancy between our findings and Kelemen et al.'s (2014) findings? First, we included 4-year-olds (12 in total, or 23% of the younger age group), whereas Kelemen and colleagues (2014) did not. Second, we probed children's understanding of evolution with minimal prompting, whereas Kelemen and colleagues (2014) used more extensive prompting. Third, we taught children evolutionary explanations using two rather lengthy training examples, whereas Kelemen and colleagues (2014) did so using only one training example. Fourth, we required children to generalize their newfound knowledge to two types of traits, whereas Kelemen and colleagues (2014) required that children generalize their knowledge to just one type. These last few points suggest that our task placed a greater cognitive load on children than did Kelemen et al.'s (2014) task—an ecologically valid load but a load nonetheless—and younger children's ability to learn evolutionary principles under such a load proved too fragile.

This fragility makes sense in light of what we know about young children's understanding of folkbiology more generally. Preschool- and early elementary school-age children tend to construe life and its properties in animistic terms (Carey, 1985; Hatano & Inagaki, 1994; Solomon & Zaitchik, 2012). They identify life with the capacity for self-directed motion (as opposed to metabolic processing), and they identify biological properties with the psychological experiences that accompany those properties (as opposed to their role in sustaining vital functions). Children of this age are still in the process of constructing a vitalist conception of bodily functions and a mechanistic conception of bodily structures, and such conceptions may need to be in place before children can learn higher-order conceptions, like evolution and natural selection.

One of the more surprising findings from our study was that learning evolutionary explanations for biological adaptation had little effect on participants' endorsement of nonevolutionary explanations, referencing the needs of individual organisms, the growth of individual organisms, or the intentions of a divine creator. These explanations were cited not only at pretest but also at posttest, often in combination with newly learned evolutionary principles, as in the following example:

> They were all born to have something different about each other [*variation*], and they all need something to use to eat [*need*]. And then they have babies, and the babies have the same traits as their parents [*inheritance*]. And then, the ones that live longer replace for the ones that don't [*population change*].

Other studies have found similar results in that people typically endorse several explanations for biological adaptation when given the opportunity to do so, some evolutionary and some not (Evans et al., 2010; Legare et al., 2013). Tensions between evolutionary and nonevolutionary explanations either go unrecognized by those who endorse both or are reconciled in terms of a synthetic model of evolution that captures aspects of both intuitive and scientific reasoning, such as a theistic model of evolution in which evolutionary change is *intentionally* and *teleologically* guided by a divine agent (Evans & Lane, 2011).

One interpretation of these findings is that participants did not learn an evolutionary schema for biological adaptation to a satisfactory degree. If they had, they should have recognized the inadequacy of nonevolutionary explanations and stopped providing them. This interpretation, though plausible from a *logical* point of view, is less plausible from a *psychological* point of view, as recent research in science education has revealed that scientific conceptions rarely replace intuitive ones. The coexistence of intuitive knowledge and scientific knowledge appears to be the rule, rather than the exception, in how people conceptualize natural phenomena (Shtulman & Harrington, 2016; Shtulman & Valcarcel, 2012).

Even adults with extensive backgrounds in science never appear to relinquish their intuitive beliefs. Under time pressure or cognitive load, they reveal nonscientific, childlike conceptions of several domains, including life (Goldberg & Thompson-Schill, 2009), matter (Potvin, Masson, Lafortune, & Cyr, 2015), motion (Foisy, Potvin, Riopel, & Masson, 2015), rational number (Vamvakoussi, Van Dooren, & Verschaffel, 2012), and physical causality (Kelemen, Rottman, & Seston, 2013). Findings of this nature suggest that the goal of science education should not be to supplant intuitive conceptions of the world with more scientific ones but to teach students when and how to discriminate between the two types of conceptions (Potvin, 2013; Shtulman & Lombrozo, in press).

Overall, our findings indicate that older elementary school-age children can be taught an evolutionary schema that (a) covers multiple evolutionary principles, (b) can be applied to multiple types of traits, and (c) can be elicited with minimal prompting, but our findings do raise several questions for further research.

First, how stable are these schemas? Kelemen et al. (2014) showed that elementary school-age children can retrieve evolutionary schemas 3 months after training, but their prompting was extensive, and it is not clear whether children would be equally successful with less prompting, as done in the present study. Second, how comprehensive are these schemas? Children in the present study were able to apply evolutionary schemas to instances of biological adaptation that differed in form and function from those used at training, but they may not be able to apply those schemas to other instances of biological change (e.g., the loss of a trait over evolutionary time or the evolution of traits in nonanimal species; Opfer, Nehm, & Ha, 2012). Third, how coherent are these schemas? As a group, the older children in our study demonstrated evidence of learning all key evolutionary principles on offer, but individual children varied in how many principles they learned, and it is not clear that a child who learned only a subset of those principles acquired a "coherent and portable relational structure," as is the goal of schema abstraction (Gentner et al., 2003, p. 402).

These questions are perhaps best pursued in the context of a longitudinal study, in which children's ability to learn evolution is assessed not just once but across several assessments periods and several tasks. That said, it should be noted that all available data—data from this study, data from Kelemen et al. (2014), and data from Legare et al. (2013)—suggest that older elementary school-age children are in fact capable of learning evolutionary ideas and should not therefore be shielded from those ideas until adolescence or even adulthood. Evolution by natural selection is foundational to the biological sciences and should thus be foundational to children's biological education as well.

Funding

This research was supported by National Science Foundation Grant No. DRL-0953384 awarded to Andrew Shtulman.

References

Berkman, M. B., & Plutzer, E. (2011, January 28). Defeating creationism in the courtroom, but not in the classroom. *Science*, *331*, 404–405. doi:10.1126/science.1198902

Berti, A. E., Toneatti, L., & Rosati, V. (2010). Children's conceptions about the origin of species: A study of Italian children's conceptions with and without instruction. *Journal of the Learning Sciences*, *19*, 506–538. doi:10.1080/10508406.2010.508027

Bishop, B., & Anderson, C. W. (1990). Student conceptions of natural selection and its role in evolution. *Journal of Research in Science Teaching*, *27*, 415–427. doi:10.1002/(ISSN)1098-2736

Blancke, S., De Smedt, J., De Cruz, H., Boudry, M., & Braeckman, J. (2012). The implications of the cognitive sciences for the relation between religion and science education: The case of evolutionary theory. *Science & Education*, *21*, 1167–1184. doi:10.1007/s11191-011-9402-z

Brem, S. K., Ranney, M., & Schindel, J. (2003). Perceived consequences of evolution: College students perceive negative personal and social impact in evolutionary theory. *Science Education*, *87*, 181–206. doi:10.1002/(ISSN)1098-237X

Brumby, M. N. (1984). Misconceptions about the concept of natural selection by medical biology students. *Science Education*, *68*, 493–503. doi:10.1002/(ISSN)1098-237X

Carey, S. (1985). *Conceptual change in childhood*. Cambridge, MA: MIT Press.

Carmichael, M. (2009, November 22). Why evolution should be taught to younger kids. *Newsweek*. Retrieved from http://www.newsweek.com/why-evolution-should-be-taught-youngerkids-76885

Catrambone, R., & Holyoak, K. J. (1989). Overcoming contextual limitations on problem-solving transfer. *Journal of Experimental Psychology: Learning, Memory, and Cognition, 15*, 1147–1156.

Christie, S., & Gentner, D. (2010). Where hypotheses come from: Learning new relations by structural alignment. *Journal of Cognition and Development, 11*, 356–373. doi:10.1080/15248371003700015

Coley, J. D., & Tanner, K. D. (2012). Common origins of diverse misconceptions: Cognitive principles and the development of biology thinking. *CBE-Life Sciences Education, 11*, 209–215. doi:10.1187/cbe.12-06-0074

Dobzhansky, T. (1973). Nothing in biology makes sense except in the light of evolution. *The American Biology Teacher, 35*, 125–129. doi:10.2307/4444260

Emmons, N. A., & Kelemen, D. A. (2015). Young children's acceptance of within-species variation: Implications for essentialism and teaching evolution. *Journal of Experimental Child Psychology, 139*, 148–160. doi:10.1016/j.jecp.2015.05.011

Evans, E. M. (2001). Cognitive and contextual factors in the emergence of diverse belief systems: Creation versus evolution. *Cognitive Psychology, 42*, 217–266. doi:10.1006/cogp.2001.0749

Evans, E. M. (2008). Conceptual change and evolutionary biology: A developmental analysis. In S. Vosniadou (Ed.), *International handbook of research on conceptual change* (pp. 263–294). New York, NY: Routledge.

Evans, E. M., & Lane, J. D. (2011). Contradictory or complementary? Creationist and evolutionist explanations of the origin(s) of species. *Human Development, 54*, 144–159. doi:10.1159/000329130

Evans, E. M., Spiegel, A. N., Gram, W., Frazier, B. N., Tare, M., Thompson, S., & Diamond, J. (2010). A conceptual guide to natural history museum visitors' understanding of evolution. *Journal of Research in Science Teaching, 47* (3), 326–353.

Foisy, L.-M., Potvin, P., Riopel, M., & Masson, S. (2015). Is inhibition involved in overcoming a common physics misconception in mechanics? *Trends in Neuroscience and Education, 4*, 26–36. doi:10.1016/j.tine.2015.03.001

Gelman, S. A. (2003). *The essential child.* Oxford, UK: Oxford University Press.

Gelman, S. A., & Rhodes, M. (2012). "Two-thousand years of stasis": How psychological essentialism impedes evolution understanding. In K. S. Rosengren, S. Brem, E. M. Evans, & G. Sinatra (Eds.), *Evolution challenges: Integrating research and practice in teaching and learning about evolution* (pp. 3–21). Cambridge, UK: Oxford University Press.

Gentner, D., Loewenstein, J., & Thompson, L. (2003). Learning and transfer: A general role for analogical encoding. *Journal of Educational Psychology, 95*, 393–408. doi:10.1037/0022-0663.95.2.393

Gick, M. L., & Holyoak, K. J. (1983). Schema induction and analogical transfer. *Cognitive Psychology, 15*, 1–38. doi:10.1016/0010-0285(83)90002-6

Goldberg, R. F., & Thompson-Schill, S. L. (2009). Developmental "roots" in mature biological knowledge. *Psychological Science, 20*, 480–487. doi:10.1111/psci.2009.20.issue-4

Gregory, T. R. (2009). Understanding natural selection: Essential concepts and common misconceptions. *Evolution: Education and Outreach, 2*, 156–175.

Hatano, G., & Inagaki, K. (1994). Young children's naive theory of biology. *Cognition, 50*, 171–188. doi:10.1016/0010-0277(94)90027-2

Johnson, S. C. (2000). The recognition of mentalistic agents in infancy. *Trends in Cognitive Sciences, 4*, 22–28. doi:10.1016/S1364-6613(99)01414-X

Kampourakis, K. (2014). *Understanding evolution.* Cambridge, UK: Cambridge University Press.

Kelemen, D. (2012). Teleological minds: How natural intuitions about agency and purpose influence learning about evolution. In K. S. Rosengren, S. Brem, E. M. Evans, & G. Sinatra (Eds.), *Evolution challenges: Integrating research and practice in teaching and learning about evolution* (pp. 66–92). Cambridge, UK: Oxford University Press.

Kelemen, D., Emmons, N. A., Schillaci, R. S., & Ganea, P. A. (2014). Young children can be taught basic natural selection using a picture-storybook intervention. *Psychological Science, 25*, 893–902. doi:10.1177/0956797613516009

Kelemen, D., Rottman, J., & Seston, R. (2013). Professional physical scientists display tenacious teleological tendencies: Purpose-based reasoning as a cognitive default. *Journal of Experimental Psychology: General, 142*, 1074–1083. doi:10.1037/a0030399

Kurtz, K. J., Miao, C.-H., & Gentner, D. (2001). Learning by analogical bootstrapping. *Journal of the Learning Sciences, 10*, 417–446. doi:10.1207/S15327809JLS1004new_2

Legare, C. H., Lane, J. D., & Evans, E. M. (2013). Anthropomorphizing science: How does it affect the development of evolutionary concepts? *Merrill-Palmer Quarterly, 59*, 168–197. doi:10.1353/mpq.2013.0009

Lombrozo, T., & Carey, S. (2006). Functional explanation and the function of explanation. *Cognition, 99*, 167–204. doi:10.1016/j.cognition.2004.12.009

Lombrozo, T., Shtulman, A., & Weisberg, M. (2006). The intelligent design controversy: Lessons from psychology and education. *Trends in Cognitive Sciences, 10*, 56–57. doi:10.1016/j.tics.2005.12.001

National Academy of Sciences. (2013). *Next generation science standards: For states, by states.* Washington, DC: National Academies Press.

Nehm, R. H., Kim, S. Y., & Sheppard, K. (2010). Academic preparation in biology and advocacy for teaching evolution: Biology versus non-biology teachers. *Science Education, 93*, 1122–1146. doi:10.1002/sce.20340

Newport, F. (2010). *Four in 10 Americans believe in strict creationism.* Princeton, NJ: Gallup Organization.

Novick, L. R., & Holyoak, K. J. (1991). Mathematical problem solving by analogy. *Journal of Experimental Psychology: Learning, Memory, and Cognition, 17*, 398–415.

Opfer, J. E., Nehm, R. H., & Ha, M. (2012). Cognitive foundations for science assessment design: Knowing what students know about evolution. *Journal of Research in Science Teaching, 49*, 744–777. doi:10.1002/tea.v49.6

Potvin, P. (2013). Proposition for improving the classical models of conceptual change based on neuroeducational evidence: Conceptual prevalence. *Neuroeducation, 1*, 16–43.

Potvin, P., Masson, S., Lafortune, S., & Cyr, G. (2015). Persistence of the intuitive conception that heavier objects sink more: A reaction time study with different levels of interference. *International Journal of Science and Mathematics Education, 13*, 21–43. doi:10.1007/s10763-014-9520-6

Samarapungavan, A., & Wiers, R. W. (1997). Children's thoughts on the origin of species: A study of explanatory coherence. *Cognitive Science, 21*, 147–177. doi:10.1207/s15516709cog2102_2

Shtulman, A. (2006). Qualitative differences between naïve and scientific theories of evolution. *Cognitive Psychology, 52*, 170–194. doi:10.1016/j.cogpsych.2005.10.001

Shtulman, A., & Calabi, P. (2012). Cognitive constraints on the understanding and acceptance of evolution. In K. S. Rosengren, S. Brem, E. M. Evans, & G. Sinatra (Eds.), *Evolution challenges: Integrating research and practice in teaching and learning about evolution* (pp. 47–65). Cambridge, UK: Oxford University Press.

Shtulman, A., & Calabi, P. (2013). Tuition vs. intuition: effects of instruction on naive theories of evolution. *Merrill-Palmer Quarterly, 59*, 141–167.

Shtulman, A., & Harrington, K. (2016). Tensions between science and intuition across the lifespan. *Topics in Cognitive Science, 8*, 118–137. doi:10.1111/tops.2016.8.issue-1

Shtulman, A., & Lombrozo, T. (in press). Bundles of contradiction: A coexistence view of conceptual change. In D. Barner & A. Baron (Eds.), *Core knowledge and conceptual change.* Oxford, UK: Oxford University Press.

Shtulman, A., & Schulz, L. (2008). The relation between essentialist beliefs and evolutionary reasoning. *Cognitive Science, 32*, 1049–1062. doi:10.1080/03640210801897864

Shtulman, A., & Valcarcel, J. (2012). Scientific knowledge suppresses but does not supplant earlier intuitions. *Cognition, 124*, 209–215. doi:10.1016/j.cognition.2012.04.005

Sinatra, G. M., Brem, S. K., & Evans, E. M. (2008). Changing minds? Implications of conceptual change for teaching and learning about biological evolution. *Evolution: Education and Outreach, 1*, 189–195.

Solomon, G. E. A., & Zaitchik, D. (2012). Folkbiology. *Wiley Interdisciplinary Reviews: Cognitive Science, 3*, 105–115. doi:10.1002/wcs.150

Son, J. Y., Smith, L. B., & Goldstone, R. L. (2011). Connecting instances to promote children's relational reasoning. *Journal of Experimental Child Psychology, 108*, 260–277. doi:10.1016/j.jecp.2010.08.011

Springer, K., & Keil, F. C. (1989). On the development of biologically specific beliefs: The case of inheritance. *Child Development, 60*, 637–648. doi:10.2307/1130729

Vamvakoussi, X., Van Dooren, W., & Verschaffel, L. (2012). Naturally biased? In search for reaction time evidence for a natural number bias in adults. *Journal of Mathematical Behavior, 31*, 344–355. doi:10.1016/j.jmathb.2012.02.001

Vasilyeva, M., & Bowers, E. (2010). Exploring the effects of similarity on mapping spatial relations. *Journal of Experimental Child Psychology, 106*, 221–239. doi:10.1016/j.jecp.2010.04.003

Ware, E. A., & Gelman, S. A. (2014). You get what you need: An examination of purpose-based inheritance reasoning in undergraduates, preschoolers, and biological experts. *Cognitive Science, 38*, 197–243. doi:10.1111/cogs.2014.38.issue-2

Using Animals to Teach Children Biology: Exploring the Use of Biological Explanations in Children's Anthropomorphic Storybooks

Megan Geerdts, Gretchen Van De Walle, and Vanessa LoBue

ABSTRACT

Research Findings: Anthropomorphism—the attribution of human characteristics to nonhuman entities—has long been a staple of children's media. However, children's experiences with anthropomorphic media may interfere with biological reasoning instead encouraging an anthropocentric view of the natural world. To date, little research has addressed the typical content of children's storybooks about animals: Do these storybooks present factual information that may support early developing biological reasoning, or do they instead focus on human-centered, psychological information that may encourage anthropocentrism? We analyzed the types of causal explanatory information that commercial storybooks about animals provide to children about 2 biological concepts that have been extensively studied in the experimental literature: biological inheritance and the transmission of illness. Using coding schemes similar to those used in prior experimental literature to assess preschool-age children's biological reasoning, we found that none of the anthropomorphized books presented children with scientifically accurate causal mechanisms. These books focused almost exclusively on social-emotional experiences as opposed to biological explanations, which may inadvertently encourage anthropocentric reasoning. *Practice or Policy*: Understanding more about the content of informal sources of early learning can help inform educators on how to best support developing knowledge about the natural world and biological properties.

Over the past 30 years, a number of researchers have investigated the origins and developmental trajectory of children's folkbiological knowledge of the natural world. One prominent view argues that children take a decidedly anthropocentric view of the natural world, conceptualizing humans and animals as distinct and extending biological properties to animals only in the context of their psychological similarity to humans (e.g., Carey, 1985). Also in line with this view, preschool-age children often fail to produce biologically specific causal explanations for biological events, such as contagion (e.g., Bibace & Walsh, 1981; Kister & Patterson, 1980) and biological inheritance (e.g., Solomon, Johnson, Zaitchik, & Carey, 1996), further suggesting that their understanding of the natural world is embedded within a folkpsychological explanatory framework. Although researchers have debated the innateness of folkbiology, different presentations of the natural world can affect children's use of either biological or psychological reasoning (e.g., Waxman, Herrmann, Woodring, & Medin, 2014). However, little work has explored the content of children's early informal experiences related to anthropocentric and biological reasoning.

In the current research we explore the kinds of evidence that children receive about the natural world and biological events that may shape the emergence of folkpsychological and folkbiological reasoning. First we discuss recent research highlighting the role that anthropomorphic media may play in encouraging anthropocentric reasoning and interfering with biological, factual learning. We present findings from a content analysis of children's animal picture books about biological properties in order to explore whether these kinds of books typically may support biological or anthropocentric reasoning. Better understanding the kind of evidence that children receive from informal learning sources, before formal schooling introduces biological explanations, can inform understanding of the developmental trajectory of young children's conceptual reasoning.

The Development of Biological Reasoning

Over the past 30 years, the nature of children's understanding of the biological world has been hotly debated (e.g., Carey, 1985; Inagaki & Hatano, 2002; Keil, 1994; Wellman & Gelman, 1998). A central focus of the argument has revolved around whether young children organize their understanding of living things within a biological or psychological framework. To assess early biological reasoning, researchers often use a biological induction task pioneered by Carey (1985). In the task, the researcher introduces an unobservable biological property (e.g., "People have spleens inside") and then asks the participant whether he or she thinks this property also applies to other animals, plants, or objects. Carey found that whereas adults reason biologically, children reason anthropocentrically: Young children readily extend properties taught on humans to other animals but make few generalizations from animals to humans or to other animals. Young children use humans as a prototypical species and extend biological properties based on behavioral similarity to humans. Over the course of early childhood, formal and informal learning experiences teach children about biological functioning and support the organization of knowledge about living things around a causally coherent and specifically biological framework (Carey, 1985; Solomon & Johnson, 2000).

In addition to research on children's general reasoning about the biological world, a large body of literature has investigated children's reasoning about specific relevant biological properties, such as contagion and biological inheritance. If children have a distinctly autonomous domain of biology, they should also demonstrate biological causal reasoning about biological properties. Research on children's developing understanding of these phenomena has yielded mixed results with respect to whether children can provide appropriate and biologically specific causal explanations for each. For instance, experimental research on children's understanding of biological inheritance often uses a cross-species adoption paradigm, in which children are presented with a baby born to one family but raised by another and are asked whether the baby will grow up to resemble the birth parents or adoptive parents on physical, behavioral, and psychological traits. For example, will a baby horse raised by cows grow up to (a) look like a horse or a cow and (b) neigh or moo? Some studies have found that it is not until at least 7 years old that children correctly favor the biological parent for physical resemblance but the adoptive parent for behavioral resemblance (e.g., Solomon et al., 1996). Young children are even willing to accept that psychological causal factors, such as maternal intentions, can influence the inheritance of both physical and psychological traits (Weissman & Kalish, 1999). Furthermore, even some children who can correctly answer prediction questions cannot correctly justify their answers and may therefore have failed to embed this knowledge in a coherent, biological causal theory (Johnson & Solomon, 1997; Williams, 2012). Similarly, children often fail to use biologically specific causal explanations of contagion-related events and are willing to attribute illness to a variety of psychological causes, such as bad behavior or wishing to be sick (Bibace & Walsh, 1981; Kister & Patterson, 1980; Perrin & Gerrity, 1981; Solomon & Cassimatis, 1999).

Children's Media and Biological Reasoning

This large body of research on children's biological reasoning has recently been extended by a growing interest in the informal learning experiences that may shape early conceptual reasoning. Cultural input is clearly important for shaping the reliance of folkbiological or folkpsychological reasoning. For instance, in communities in which alternative frameworks for illness (i.e., different from Western biomedical frameworks) are available, both children and adults endorse culturally specific alternative beliefs, such as supernatural causal mechanisms (e.g., Inagaki & Hatano, 2002; Legare, Evans, Rosengren, & Harris, 2012; Legare & Gelman, 2008). What kinds of informal sources may encourage young children to develop anthropocentric or biological reasoning about the natural world before exposure to formal schooling?

One source that has been the focus of a growing body of research is children's picture books. Nearly all (91%) children in the United States younger than the age of 5 years have been exposed to storybooks, with the average age of first exposure at 5 months (Common Sense Media, 2011). Furthermore, 83% of children younger than the age of 6 years read or are read to each day for an average of 48 min, regardless of family income (Rideout & Hammel, 2006). Thus, storybook reading is a prevalent experience in young children's daily lives that may serve as an important source of cultural learning before formal schooling.

Recent research has shown that preschool-age children can learn about the natural world and biological processes from storybooks and that the presentation of information can affect children's reasoning about animals (e.g., Ganea, Ma, & DeLoache, 2011; Geerdts, Van de Walle, & LoBue, 2015b; Kelemen, Emmons, Seston Schillaci, & Ganea, 2014; Legare, Lane, & Evans, 2013). Ganea et al. (2011) found that even children as young as 4 years old were able to generalize information about color camouflage from storybook animals to real animals. Storybook interventions can also be used to teach young children more complex biological properties: 5- to 8-year-old children read scientifically accurate natural selection narratives exhibited significant gains in knowledge about evolution, especially in the youngest age group (Kelemen et al., 2014). Thus, storybooks can serve to increase children's biological knowledge.

Of particular relevance to how children learn about the biological world, children's storybooks are full of anthropomorphic animals. An examination of more than 1,000 modern picture books found that nearly half featured animals as significant characters, with only a quarter of those animals featured in natural settings. However, even when portrayed in their natural settings, these animals were typically anthropomorphized with names, communicative abilities, and recognizably human behavior (Marriott, 2002). However, storybooks that portray animals anthropomorphically may interfere with biological learning and further encourage anthropocentric reasoning. Preschool-age children read anthropomorphic storybooks are less likely to use biological reasoning (Waxman et al., 2014) and less likely to learn factual information (Ganea, Canfield, Simons-Ghafari, & Chou, 2014). Waxman et al. (2014) read 5-year-old children either a realistic animal storybook (*First Animal Encyclopedia*) or an anthropomorphic animal storybook (the Berenstain Bears) before completing the biological induction task (Carey, 1985). Children read the realistic storybook exhibited a biological pattern of responses, attributing new biological properties between animals and humans equally. However, children of the same age who were read an anthropomorphic storybook exhibited the typical anthropomorphic, human-centered pattern of reasoning. Thus, anthropomorphic storybooks may support human-centered reasoning, whereas realistic storybooks may encourage biological reasoning. Similarly, Ganea et al. (2014) found that anthropomorphic storybooks, especially those with anthropomorphic pictures, do affect children's willingness to attribute psychological properties to animals. For example, 3-, 4-, and 5-year-old children who were read an anthropomorphic storybook about an unfamiliar animal were more likely to generalize psychological properties to real animals and, for younger children especially, were less likely to remember novel facts about the animals than children read a factual storybook. Similarly, 5- to 12-year-old children read stories with anthropomorphic explanations about evolutionary change ("Brownbirds like to have different beaks because they want to eat different sorts of food") were more likely to endorse desire-based, anthropomorphic evolutionary explanations than those read stories with need-based or natural selection explanations (Legare et al., 2013).

Although this growing body of research suggests that factual representations may be more beneficial for supporting biological reasoning, a confounding factor is the kind of information presented within these storybooks. Waxman et al. (2014) compared children's reasoning after reading one of two books about bears that differed in terms of the realism of the animals, but the books also differed in the kind of information highlighted. The anthropomorphic book focused on human-specific events related to social-emotional issues, like the trouble with getting children to stick to bedtime routines, whereas the factual book presented information such as the habitat and diet of bears. In other studies that have more closely controlled the kind of information, children did not show substantial differences in biological reasoning and in fact learned just as much about biological properties from anthropomorphic books (Ganea et al., 2011; Geerdts et al., 2015b). For instance, in a recent study (Geerdts et al., 2015b), 4- and 5-year-old children were read a book about color camouflage that presented factual, biological explanations but used either realistic or anthropomorphic language and pictures. Here anthropomorphism did not decrease factual learning: Children were not more likely to explain camouflage using psychological, intentional explanations. An important question then is what kinds of explanatory information about biological properties do commercial anthropomorphic storybooks provide?

The Current Research

In the current study, we ask whether storybooks with animal characters provide causal explanatory information about these biological properties. To the extent that they do provide causal explanations, are these explanations biological or psychological in nature? Thus, the major goal of this study is to examine whether children's biological storybooks with animals provide a potential source of information to support the development of sophisticated biological knowledge. Storybooks about cross-species adoption, for example, may attribute physical resemblance to biological parents and behavioral resemblance to adopted parents. In addition, they may explain why the inheritance of physical and behavioral properties differs by contrasting the role of inheritance or origins with that of shared experiences. However, they may also fail to provide such information. The use of anthropomorphism in particular may lead readers to focus on psychological mechanisms typical of humans, such as family bonds and emotional ties. In the current study, we are interested in examining whether storybooks provide children with the specific type of information they would need to succeed on experimental tasks used to assess children's biological causal knowledge. Thus, we create our coding schemes by adapting schemes used in prior research to code children's explanations for these biological phenomena. We may find that the types of explanations commonly present in these storybooks resemble the explanations used by children in previous research.

Informal learning experiences, such as picture book reading, may be especially relevant sources of biological information for preschoolers. As children begin formal schooling, they are taught factual and causal biological knowledge. For instance, by first grade, formal instruction about germs is part of the Common Core State Standards in many states (California Department of Education, 2011). Potential informal, cultural sources of biological reasoning, like picture books, are most relevant prior to formal schooling. Thus, in the present study, we focus on storybooks marketed to a preschool audience.

Methods

Storybook Selection

One of the largest online book retailers, Amazon.com, was used to compile a list of storybooks in June 2012. Broad keywords (*adoption, illness*, and *contagion*) were used to produce an initial sample. We then applied limiters to restrict our search to English-language children's storybooks for preschoolers (ages 3–6 years). Storybooks were scanned for content to be sure that they fit our age and language limiters, were designed for a general audience (i.e., we excluded religious storybooks, storybooks created specifically for adopted children, and storybooks that aimed to help children overcome their fear of the doctor), and featured animal main characters. The result was a sample of 20 adoption books and 25 illness books (see the Appendix for a full list of titles and authors).

Coding

All of the storybooks were analyzed and coded for (a) the use of anthropomorphism and (b) explanations/mechanisms regarding each biological property.

Anthropomorphism

Both the language and pictures in each book were coded for the use of anthropomorphism. *Anthropomorphism* is defined here as the attribution of human-specific characteristics (e.g., motivations, behaviors, actions) to nonhuman animals. Each book was placed into one of four categories based on how animals were depicted throughout the story:

(1) Realistic pictures and realistic language
(2) Realistic pictures and anthropomorphic language
(3) Anthropomorphic pictures and realistic language
(4) Anthropomorphic pictures and anthropomorphic language

Books were coded as having anthropomorphic pictures if the animals were depicted engaging in any human-specific behaviors or possessing human-specific capabilities (e.g., living in a house, reading, walking upright, wearing clothing). Books that depicted animals engaging in species-typical behaviors and actions were coded as realistic. Books were coded as using anthropomorphic language if intentional language was used to describe the animals' behavior (Ganea et al., 2011) or if the animals were ever depicted as speaking or thinking, either explicitly through direct speech or implicitly through narration. Books were coded as using realistic language if they used only neutral, factual language to describe or explain the animals' behavior without implying underlying mental states or human-specific capabilities. Two coders (the first author and an undergraduate research assistant) independently coded all of the storybooks. Interrater reliability between the two raters indicated very good agreement (Cohen's $\kappa = .809$). A third trained coder reviewed both coders' selections, and any disagreements were resolved through discussion between the third and first coders.

Explanations for Biological Properties

The text of each storybook was also coded for reference to biological and psychological explanations and/or mechanisms regarding each biological property. The explanations and mechanisms were selected on the basis of prior research that investigated children's own explanations for the biological properties.

Biological Inheritance. The text of each biological inheritance storybook was coded for reference to specific physical and behavioral similarities and differences between offspring and their biological and adoptive parents. These coding definitions were derived from prior experimental research investigating children's domain-specific causal understandings of inheritance (Johnson & Solomon, 1997; Solomon et al., 1996; Williams, 2012; Williams & Smith, 2006). Our coding scheme focused on possible sources for the offspring's physical (i.e., bodily characteristics) and nonphysical (i.e., beliefs, emotions, preferences, activities) traits. Comments about specific traits were classified according to the context within which they were described using the following categories:

- Inheritance: the passing of traits from parents to offspring via biological inheritance or biological kinship (e.g., "Jimmy got his blue eyes from his parents' genes" or "Susan inherited her mom's love of traveling")
- Adoptive family resemblance: a general statement that either compared or contrasted adoptive family members regarding resemblance on specific properties without providing a possible mechanism for resemblance (e.g., "From the beginning, Jack was much bigger than his sisters")

- Biological family resemblance: a general statement that either compared or contrasted biological family members regarding resemblance on specific properties without providing a possible mechanism for resemblance (e.g., "Pink's brother and sister piglets were strong and healthy, but he was tiny and weak")
- Generic references to category membership: generic statements that appealed to the individual's membership in a specific category to explain the possession of a specific trait (e.g., "Cats drink milk")

In addition, for nonphysical properties, the following two codes were applied:

- Shared social activities (adoptive): a social activity shared with the adoptive family members (e.g., "The three of them walked home and talked about all of the things they would do as a family")
- Shared social activities (biological): a social activity shared with the biological family members (e.g., "They liked to wrestle and nip at each other")

Contagion. For the illness stories, the coding scheme focused on possible mechanisms for illness causation. The scholarly literature on children's understanding of illness and contagion generally focuses on whether children conceptualize illness as caused by a biological mechanism (germs) or a psychological mechanism (punishment for bad behavior, desires). We coded the illness stories based on the presence of the following possible explanations (regardless of scientific accuracy) for why the character(s) became sick:

- Germs: These stories explicitly mentioned germs, viruses, bacteria, or any other medical origin for the illness.
- Physical proximity: This code was used if the story either explicitly or implicitly mentioned physical proximity as a possible cause; for instance, a story in which a character took care of one character who was sick and then himself or herself became sick with no reference to any other mechanism was coded as providing a physical proximity mechanism.
- Behavioral: This code was used for any specific activity or behavior that led to the child getting sick, such as not wearing a coat or eating too much junk food.
- Psychological: This code was used for any psychological or magical cause that led to getting sick, such as wishing to be sick, witchcraft, or punishment for behaving badly.
- Unspecific/irrelevant: These stories simply described the illness event with no specific mention of any cause.

Each storybook was coded binomially as either having made reference or not having made reference at any point in the storybook to each possible category of information. Thus, one storybook may have received more than one code. Two coders (the first author and an undergraduate research assistant) independently coded all of the titles. Interrater reliability between the two raters for the adoption stories was found to be moderate (Cohen's κ = .55), and for the illness stories interrater reliability was found to be substantial (Cohen's κ = .74). A third trained coder reviewed both coders' selections, and any disagreements were resolved through discussion.

Results

Anthropomorphism

First we explored the use of anthropomorphic language and images in the adoption and contagion storybooks. Results are summarized in Table 1. None of the cross-species adoption storybooks used realistic or factual language; all used anthropomorphic language to describe the animals as possessing complex feelings, emotions, and behaviors and engaging in social interactions. However, it was not necessarily the case that anthropomorphic language was always accompanied by anthropomorphic images: Close to a third of the storybooks (30%) portrayed the animals visually using realistic drawings or photographs. Many of the storybooks showed animals in their natural habitat behaving realistically despite using narration to tell a very human, emotional story. However, this may have been due more to the nature of the story than a desire to represent animals accurately. To make the theme of an individual being adopted into a different family as obvious as possible, it may be easiest to maintain the visual and behavioral differences between the species rather than portraying them all visually as humans.

Similarly, none of the illness storybooks used realistic, factual language; rather, all used anthropomorphic language. In addition, nearly all of the illness books (96%) used anthropomorphic drawings. Illness books were significantly more likely to use anthropomorphic drawings than the adoption books, $\chi^2(1, N = 45) = 5.72, p = .017$. The only illness book that did not use anthropomorphic drawings was one that told the true story of a sick sea turtle that was found, rescued, cured, and released using real photographs. However, this story still used heavily anthropomorphic language to describe how the sea turtle, named Caroline, was feeling and what she was thinking as she recovered from the "turtle flu" (Rathmell, 2005, p. 4).

Explanations for Biological Properties

Biological Inheritance

The adoption storybooks were coded for reference to specific physical and behavioral similarities and differences between offspring and their biological and adoptive parents. Results are presented in Table 2. Overall, storybooks were more likely to focus on behavioral than physical properties: All books made at least one reference to behavioral properties, whereas only 60% of the books referred to physical properties, $\chi^2(1, N = 40) = 7.66, p = .006$.

For physical properties, none of the storybooks used inheritance, the correct biological mechanism, to explain the transmission of physical properties between parents and their biological offspring. About a third of the books provided biological family resemblance (30%) and/or category membership (30%) as possible sources for the child's physical properties. Given that biological families should be more physically similar than adoptive families, we were interested in seeing whether explicit comparisons were more often made between biological and adoptive families. Chi-square tests revealed that the books were no more likely to draw physical property comparisons between the child and his or her biological family than between the child and his or her adoptive family, $\chi^2(1, N = 40) = 0.94, p = .33$. Thus, storybooks did not provide any

Table 1. Number (%) of Storybooks With Anthropomorphic Language and/or Pictures.

Type of Book	Realistic Pictures, Realistic Language	Realistic Pictures, Anthropomorphic Language	Anthropomorphic Pictures, Realistic Language	Anthropomorphic Pictures, Anthropomorphic Language
Adoption books	0	6 (30%)	0	14 (70%)
Illness books	0	1 (4%)	0	24 (96%)

Table 2. Number (%) of Biological Inheritance Books That Were Coded as Having Made Reference at Any Point in the Storybook to Each Possible Category of Information.

Property	Inheritance	Adopted Family Resemblance	Biological Family Resemblance	Generic Category Reference	Shared Social Activities (Adopted)	Shared Social Activities (Biological)
Physical	0 (0%)	10 (50%)	6 (30%)	6 (30%)	-	-
Nonphysical	0 (0%)	7 (35%)	0 (0%)	9 (45%)	14 (70%)	2 (10%)

information about biological mechanisms for physical inheritance but instead focused simply on comparing and contrasting physical properties between different family members and/or animal kinds.

For behavioral properties, similarly, none of the storybooks referenced inheritance in explaining the possession of specific behavioral properties. As with physical properties, some books provided adoptive family resemblance (35%) and/or generic category references (45%) as possible sources for the child's behavioral properties. Most frequently, the books referred to social activities that were shared with the adoptive family (70%). Little behavioral information was given about the biological family at all: No books referred to biological family resemblance for behavioral properties, and only two books (10%) made reference to shared social activities with the biological family. As adopted children should share more similarities with adopted parents in terms of behavioral properties, we compared references to biological and adoptive families for behavioral and social properties. It was significantly more likely for books to contain references to the adoptive family than the biological family in both behavioral resemblance and shared social activities: behavioral resemblance, $\chi^2(1, N = 40) = 6.23, p = .013$; shared social activities, $\chi^2(1, N = 40) = 12.60, p = .004$.

Contagion

Here we present results for biological and psychological explanatory information found in the illness storybooks. Results are presented in Table 3. Like the adoption storybooks, most illness storybooks (88%) failed to provide the correct, biologically specific causal mechanism underlying the phenomenon of contagion. No storybooks specifically mentioned germs, but two did mention medical causes, including tonsils and needing an operation to "fix a stripe that slipped out of place" (Janosch, 1985, p. 14). Many stories implied that physical proximity contributed to the transmission of illness (36%), but none of these referenced actual mechanisms of transmission such as germs. In many stories, a character got sick and then the character's caregiver soon became sick, leading to a role reversal in the care relationship. These stories did not mention why illness is contagious or how it passes between individuals but provided evidence that physical proximity may be a possible factor. Few stories mentioned psychological (8%) or behavioral (12%) causes of illness. Often stories did not mention any possible mechanisms for illness (36%) and simply focused on the experience of being sick.

Discussion

A large body of previous research has focused on the content and development of young children's early understanding of the biological world (e.g., Carey, 1985; Inagaki & Hatano, 2002; Wellman & Gelman, 1998). However, little previous research furnishes data from which one can draw conclusions about the information that is present in children's everyday experiences and how this input may contribute to children's knowledge development. In the current research, we focused on one specific and common cultural influence that may affect the development of biological and anthropocentric reasoning: children's storybooks.

We found that media depicting animals engaging in a biological event often present a psychologically motivated story. The adoption and illness storybooks almost always portrayed animals anthropomorphically in both pictures and language. These results are consistent with prior literature reporting that animals in children's media are largely portrayed as human-like, with a focus on social and emotional traits (e.g., Marriott, 2002).

In addition, these storybooks do not appear to be a potential source through which children could learn factual, biologically specific causal mechanisms, in that they rarely referred to biological inheritance or

Table 3. Number (%) of Illness Books That Were Coded as Having Made Reference at Any Point in the Storybook to Each Possible Category of Information.

Germs	Physical Proximity	Behavioral	Psychological	Unspecific/Irrelevant
3 (12%)	9 (36%)	3 (12%)	2 (8%)	9 (36%)

germs. Stories about cross-species adoption never referenced biological inheritance in explaining physical differences between an adopted child and an adoptive family of a different species. The storybooks did often highlight shared social experiences with the adoptive family, which may help children understand that these environmental factors are important for the transmission of behavioral properties. However, they did not do the same for physical properties: Although the storybooks often compared and contrasted the child and the adoptive parents, they rarely referenced biological parentage as a possible source of physical properties. The common overall theme of these books was that family is defined not by physical or genetic commonalities but by emotional and social connections, with no mention made of the role of biological parentage in the adoptee's possession of specific physical properties.

Similarly, illness stories focused on the physical and emotional consequences of being sick and often presented physical proximity as an important factor in predicting contagion. Given our findings of the kind of input children may receive in typical informal learning experiences, it is not very surprising that children themselves do not always reason according to specifically biological causal mechanisms in empirical adoption and contagion paradigm tasks. It is interesting that illness storybooks rarely use the kinds of psychological causal explanations that some previous research has found children often provide themselves, such as citing illness as punishment for doing something wrong or immoral (Bibace & Walsh, 1981; Kister & Patterson, 1980). Thus, the animals in both sets of storybooks we analyzed largely taught about important emotional and relational properties rather than explaining the biological facts important for understanding these properties.

These findings have important methodological and theoretical implications. Empirical studies frequently use the cross-species adoption paradigm to assess children's naïve biology (e.g., Johnson & Solomon, 1997; Solomon et al., 1996; Springer, 1996; Springer & Keil, 1991). This paradigm pits social against biological influences on property transfer from parent to child. However, the current work examining the information presented to children via published books suggests that from this common childhood media activity children are not receiving input that reflects a clear differentiation of biological and social sources for physical and behavioral properties. Storybooks are much more apt to focus on social influences and properties, neglecting the role of biological parents in physical property transmission. Furthermore, children were never provided with explicit causal information such as biological inheritance. Thus, it is perhaps not surprising that many children have failed to provide biologically relevant responses in the empirical literature (e.g., Bibace & Walsh, 1981; Johnson & Solomon, 1997; Kister & Patterson, 1980; Solomon et al., 1996), as it is not clear that they are receiving the relevant information from at least one of their everyday experiences.

Theoretically speaking, this work has implications for the origins of children's anthropocentric reasoning bias. Our research is consistent with the proposal that the development of anthropocentrism is mediated by experiences (e.g., Herrmann, Waxman, & Medin, 2010; Waxman & Medin, 2007). Specifically, Waxman and colleagues (2007, 2014) have postulated that children raised in urban environments are exposed to anthropomorphized animals through media and cultural discourse and that these experiences lead children to adopt an anthropocentric view of the natural world. Although the current work was not experimental and therefore provides no causal evidence concerning the direction of the relationship between media and knowledge, it does demonstrate that at least one form of media to which children are widely and frequently exposed from a very young age—storybooks—almost exclusively promotes such an anthropocentric bias. The storybooks we analyzed highlighted psychological and behavioral characteristics of animals, simultaneously omitting information on biological properties of animals or on biologically relevant aspects of adoption or contagion. Although both urban and rural children may be similarly exposed to such media, urban children may be less likely to have other, more biologically accurate experiences of animals to counter such anthropomorphized media portrayals. Prior experimental research suggests that anthropomorphism interferes with the developmental of biological reasoning and factual learning (Ganea et al., 2014; Legare et al., 2013; Waxman et al., 2014), suggesting that much of children's informal literature on the biological properties focused on in the present study may not be supporting biological learning at all and may in fact be contributing to the development of anthropocentric reasoning.

This study was limited in scope because we only focused on one source of potential information for preschoolers—picture books. Specifically, we chose to narrowly focus on storybooks with animals to see whether biological explanations may be used in conjunction with anthropomorphic depictions of the natural world. It is entirely possible that books with human characters contain different kinds of explanations and may be more apt to focus on biological explanations than the books we reviewed here. In addition, children certainly gather information about biological properties from other sources, including other forms of media, parents, siblings, and firsthand experiences like being sick. However, other research suggests that the findings reported here are highly representative of other typical experiences preschool-age children have that have to do with animals and biological concepts. For instance, children's interactions with parents in book reading (Geerdts, Kim, & Van De Walle, 2013) and in zoos and science centers (Geerdts, Van de Walle, & LoBue, 2015c), interactions with teachers in preschool settings (Kallery & Psillos, 2004; Thulin & Pramling, 2009), and even direct interactions with companion pets (Geerdts, Van de Walle, & LoBue, 2015a) are highly likely to contain anthropomorphic information. In addition, during storybook reading about cross-species adoption, parents provide information that is very similar to the storybooks we analyzed in the current study, with no focus on biological inheritance and a substantial focus on shared social experiences that tie families together (Geerdts et al., 2013). Together with the current research, these observations of parents and their children provide further support for the conclusion that children (especially from urban, majority populations) are exposed to anthropocentric biases through both media and discourse.

To date, a large amount of research has focused on children's developing understanding of biological concepts for a largely theoretical purpose: to better understand the origins and development of children's naïve domains of knowledge. Conceptual tasks have been focused on understanding when children understand biological and psychology as separate domains of reasoning abut the natural world. However, little research has focused on the everyday experiences and larger cultural influences, such as media, that impact the developmental trajectory of children's biological concepts. Future work should continue to combine experimental with more descriptive and naturalistic methodologies, characterizing real-world experiences so that researchers can later measure learning outcomes experimentally. Such methodologies can help experts gain a better understanding of how direct and indirect daily experiences can optimally support early biological knowledge development. For educators specifically, research on early sources of informal learning can shed light on the emerging understanding of biology that children bring to the classroom. Better understanding the kinds of information that children are receiving outside the classroom can highlight the areas that educators can focus on to optimally support early biological knowledge development. Our research shows that many of children's everyday interactions with biological concepts are not providing causal explanatory information, which may be an area that early education teachers can target with interventions. Additional research is also needed to provide recommendations for educators about the kinds of books, anthropomorphic or factual, that support factual biological knowledge development while still engaging children in learning about the natural world.

Acknowledgments

Megan Geerdts is now in the Department of Undergraduate Psychology at Chestnut Hill College.

ORCID

Megan Geerdts http://0000-0003-4398-0128

References

Bibace, R., & Walsh, M. E. (1981). Children's conceptions of illness. *New Directions for Child and Adolescent Development, 1981*(14), 31–48. doi:10.1002/cd.23219811404

California Department of Education. (2011). *A look at kindergarten in California public schools and the common core state standards.* Sacramento, CA: Department of Education.

Carey, S. (1985). *Conceptual change in childhood.* Cambridge, MA: MIT Press.

Common Sense Media. (2011). *Zero to eight: Children's media use in America.* Retrieved from www.commonsensemedia.org

Ganea, P. A., Canfield, C. F., Simons-Ghafari, K., & Chou, T. (2014). Do cavies talk? The effect of anthropomorphic picture books on children's knowledge about animals. *Frontiers in Psychology, 5*, 283. doi:10.3389/fpsyg.2014.00283

Ganea, P. A., Ma, L., & DeLoache, J. S. (2011). Young children's learning and transfer of biological information from picture books to real animals. *Child Development, 82*, 1421–1433. doi:10.1111/j.1467-8624.2011.01612.x

Geerdts, M., Kim, V., & Van De Walle, G. (2013, April). *Biological explanations in parent-child storybook reading.* Poster presented at the biennial meeting of the Society for Research in Child Development, Seattle, WA.

Geerdts, M., Van De Walle, G. A., & LoBue, V. (2015a). Daily animal exposure and children's biological concepts. *Journal of Experimental Child Psychology, 130*, 132–146. doi:10.1016/j.jecp.2014.10.001

Geerdts, M., Van De Walle, G. A., & LoBue, V. (2015b). Learning about real animals from anthropomorphic media. *Journal of Imagination, Cognition, and Personality.* Advance online publication. doi:10.1177/0276236615611798

Geerdts, M., Van De Walle, G. A., & LoBue, V. (2015c). Parent–child conversations about animals in informal learning environments. *Visitor Studies, 18*(1), 39–63. doi:10.1080/10645578.2015.1016366

Herrmann, P., Waxman, S. R., & Medin, D. L. (2010). Anthropocentrism is not the first step in children's reasoning about the natural world. *Proceedings of the National Academy of Sciences, USA, 107*(22), 9979–9984. doi:10.1073/pnas.1004440107

Inagaki, K., & Hatano, G. (2002). *Young children's naïve thinking about the biological world.* New York, NY: Psychology Press. doi:10.4324/9780203759844

Janosch. (1985). *"I'll make you well, Tiger," said the Bear.* New York: NY: Adama Books.

Johnson, S. C., & Solomon, G. E. (1997). Why dogs have puppies and cats have kittens: The role of birth in young children's understanding of biological origins. *Child Development, 68*(3), 404–419. doi:10.2307/1131668

Kallery, M., & Psillos, D. (2004). Anthropomorphism and animism in early years science: Why teachers use them, how they conceptualise them and what are their views on their use. *Research in Science Education, 34*, 291–311. doi:10.1023/B:RISE.0000044613.64634.03

Keil, F. C. (1994). The birth and nurturance of concepts by domains: The origins of concepts of living things. In L. A. Hirschfeld & S. A. Gelman (Eds.), *Mapping the mind: Domain specificity in cognition and culture*, pp. 234-254. New York, NY: Cambridge University Press. doi:10.1017/CBO9780511752902

Kelemen, D., Emmons, N. A., Seston Schillaci, R., & Ganea, P. A. (2014). Young children can be taught basic natural selection using a picture-storybook intervention. *Psychological Science, 25*(4), 893–902. doi:10.1177/0956797613516009

Kister, M. C., & Patterson, C. J. (1980). Children's conceptions of the causes of illness: Understanding of contagion and use of immanent justice. *Child Development, 51*, 839–846. doi:10.2307/1129472

Legare, C. H., Evans, E. M., Rosengren, K. S., & Harris, P. L. (2012). The coexistence of natural and supernatural explanations across cultures and development. *Child Development, 83*, 779–793. doi:10.1111/j.1467-8624.2012.01743.x

Legare, C. H., & Gelman, S. A. (2008). Bewitchment, biology, or both: The co-existence of natural and supernatural explanatory frameworks across development. *Cognitive Science, 32*, 607–642. doi:10.1080/03640210802066766

Legare, C. H., Lane, J., & Evans, E. M. (2013). Anthropomorphizing science: How does it affect the development of evolutionary concepts? *Merrill-Palmer Quarterly, 59*(2), 168–197. doi:10.1353/mpq.2013.0009

Marriott, S. (2002). Red in tooth and claw? Images of nature in modern picture books. *Children's Literature in Education, 33*(3), 175–183. doi:10.1023/A:1019677931406

Perrin, E. C., & Gerrity, P. S. (1981). There's a demon in your belly: Children's understanding of illness. *Pediatrics, 67*, 841–849.

Rathmell, D. (2005). *Carolina's story: Sea turtles get sick too! Mt.* Pleasant, SC: Sylvan Dell Publishing.

Rideout, V., & Hammel, E. (2006). *The media family: Electronic media in the lives of infants, toddlers, preschoolers and their parents.* Menlo Park, CA: Kaiser Family Foundation.

Solomon, G. E. A., & Cassimatis, N. L. (1999). On facts and conceptual systems: Young children's integration of the understandings of germs and contagion. *Developmental Psychology, 35*, 113–126. doi:10.1037/0012-1649.35.1.113

Solomon, G. E. A., & Johnson, S. C. (2000). Conceptual change in the classroom: Teaching young children to understand biological inheritance. *British Journal of Developmental Psychology, 18*, 81–96. doi:10.1348/026151000165580

Solomon, G. E. A., Johnson, S. C., Zaitchik, D., & Carey, S. (1996). Like father, like son: Young children's understanding of how and why offspring resemble their parents. *Child Development, 67*, 151–171. doi:10.2307/1131693

Springer, K. (1996). Young children's understanding of a biological basis for parent–offspring relations. *Child Development, 67*, 2841–2856. doi:10.2307/1131755

Springer, K., & Keil, F. C. (1991). Early differentiation of causal mechanisms appropriate to biological and non-biological kinds. *Child Development*, *62*(4), 767–781. doi:10.2307/1131176

Thulin, S., & Pramling, N. (2009). Anthropomorphically speaking: On communication between teachers and children in early childhood biology education. *International Journal of Early Years Education*, *17*(2), 137–150. doi:10.1080/09669760902982331

Waxman, S. R., Herrmann, P., Woodring, J., & Medin, D. (2014). Humans (really) are animals: Picture-book reading influences five-year-old urban children's construal of the relation between humans and non-human animals. *Frontiers in Psychology*, *5*, 172. doi:10.3389/fpsyg.2014.00172

Waxman, S., & Medin, D. (2007). Experience and cultural models matter: Placing firm limits on childhood anthropocentrism. *Human Development*, *50*, 23–30. doi:10.1159/000097681

Weissman, M. D., & Kalish, C. (1999). The inheritance of desired characteristics: Children's view of the role of intention in parent-offspring resemblance. *Quarterly Journal of Experimental Child Psychology*, *73*, 245–265. doi:10.1006/jecp.1999.2505

Wellman, H., & Gelman, S. (1998). Knowledge acquisition in foundational domains. In D. Kuhn & R. S. Siegler (Eds.), *Handbook of child psychology* (pp. 523–574). New York, NY: Wiley.

Williams, J. M. (2012). Children and adolescents' understandings of family resemblance: A study of naïve inheritance concepts. *British Journal of Developmental Psychology*, *30*(2), 225–252. doi:10.1111/j.2044-835x.2011.02031.x

Williams, J. M., & Smith, L. A. (2006). Social and experiential influences on the development of inheritance concepts. *International Journal of Behavioral Development*, *30*(2), 148–157. doi:10.1177/0165025406063630

Appendix

List of Titles and Authors of Analyzed Children's Books`

Adoption Storybooks

A Mother for Choco by Keiko Kasza
Little Miss Spider by David Kirk
Guji Guji by Chih-Yuan Chen
A Mama for Owen by Marion Dane Bauer and John Butler
Our Twitchy by Kes Gray and Mary McQuillan
Little Pink Pup by Johanna Kerby
Nikolai, the Only Bear by Barbara Joosse and Renata Liwska
Mrs. Hen's Big Surprise by Christel Desmoinaux
Cora and the Elephants (Picture Puffins) by Lissa Rovetch
You're Somebody Special, Walliwigs by Joan Rankin
The Little Green Goose by Adele Sansone and Faust Anke
Owen and Mzee by Michelle Y. Glennon
Bullfrog and Gertrude Go Camping by Rosamond Dauer
Goose by Molly Bang
Honey Badgers by Jamison Odone
Paddington Bear by Michael Bond and R. W. Alley
The Nanny Goat's Kid by Jeanne Willis and Tony Ross
Giant Jack by Birte Muller
All Together Now by Anita Jeram
Stellaluna by Janell Cannon

Illness Storybooks
Ah-choo! Lion's Got the Flu by Hagino Chinatsu
Bear Feels Sick by Karma Wilson and Jane Chapman
The Berenstain Bears: Sick Days by Jan Berenstain and Mike Berenstain
Carolina's Story: Sea Turtles Get Sick Too! by Donna Rathmell and Barbara J. Bergwerf
Doctor Maisy by Lucy Cousins
Don't You Feel Well, Sam? by Amy Hest and Anita Jeram
Dr. Duck by H. M. Ehrlich and Laura Rader
Feel Better, Ernest! by Gabriel Vincent
Felix Feels Better by Rosemary Wells
How Do Dinosaurs Get Well Soon? by Jane Yolen and Mark Teague
I Wish I Was Sick, Too! by Franz Brandenberg and Aliki
I'll Make You Well, Tiger, Said the Bear by Janosch
Little Brown Bear Is Sick by Claude Lebrun and Daniele Bour
Little Raccoon Catches a Cold by Susan Canizares and Christopher Denise
Llama Llama Home With Mama by Anna Dewdney
Marty Moose Gets Sick by Kiki

Miss Bindergarten Stays Home from Kindergarten by Joseph Slate and Ashley Wolff
Morris Has a Cold by Bernard Wiseman
Pigs Make Me Sneeze! by Mo Willems
Sam and Violet's Get Well Book by Nicole Rubel
Taking Care of Mama by Mitra Modarressi
Teddy Bears Cure a Cold by Susanna Gretz and Alison Sage
The Sick-in-Bed Birthday by Linda Wagner Tyler
The Sniffles for Bear by Bonny Becker and Kady MacDonald Denton
Who's Sick Today? by Lynne Cherry

Body and soul: Do children distinguish between foods when generalizing biological and psychological properties?

Jean-Pierre Thibaut, Simone P. Nguyen, and Gregory L. Murphy

ABSTRACT

Research Findings: In 2 experiments, we tested whether children generalize psychological and biological properties to novel foods. We used an induction task in which a property (either biological or psychological) was associated with a target food. Children were then asked whether a taxonomically related and a script-related food would also have the property. In a yes/no task (Experiment 1) 9-year-olds preferentially generalized the property to taxonomically related foods, but 4- and 6-year-olds did not. In a forced-choice task (Experiment 2; 4- to 6-year-olds), children preferred the taxonomic choice over the script choice. This preference was weak at age 4 but established by age 5. In both experiments, and all age groups, biological properties, and psychological properties were treated similarly. It is argued that the children do not distinguish biological and psychological properties of food most likely because they believe that psychological properties are caused by biological dispositions. *Practice or Policy*: We argue that nutrition education should take advantage of children's existing knowledge of food categories and how children generalize knowledge from 1 food to another. In particular, children have good knowledge of taxonomic categories and can best access that knowledge when they are required to compare different foods.

One way that children can learn about the world is through category-based induction, the generalization of information from a known to an unknown category member. There is a large body of research documenting children's ability to use a premise category to make inferences about the properties of unfamiliar instances of that category (for a review, see Gelman & Davidson, 2013; Hayes, 2007; Murphy, 2002). For example, if a child learns a new fact about a tabby cat, the child may reasonably conclude based on category membership that the fact is also true of a Siamese cat.

The present research examines children's inductive inferences with food categories. Children's ability to use food categories for inferences about nutrition, and specifically the health-related effects of eating, carries both theoretical and practical importance. Theoretically speaking, children's behavior in induction tasks using food categories is related to developments in their naive theory of biology. A substantial body of research has documented children's naive theory of biology, including an early distinction between natural kinds and artifacts and an understanding of numerous biological processes (for a review, see Gelman & Kalish, 2006; Inagaki & Hatano, 2006; Wellman & Gelman, 1997). Past research has found evidence for increases in reasoning about some foods as natural kinds (e.g., fruits, vegetables) between preschool and second grade (Gelman, 1988). A recent study also found that 3- and 4-year-olds have knowledge of the origins of natural versus processed foods (e.g., orange vs. cookie), recognizing that the former foods grow outside whereas the latter are made in factories (Girgis & Nguyen, 2015).

Many biological processes that have been studied within the context of children's naive theory of biology are highly relevant to nutritional issues, such as growth, illness, and organ functioning. Studying children's use of food categories for induction addresses how children acquire new information about nutrition and how this information is integrated into a broader biological framework that includes other biological processes. For example, if children believe that salmon is good for you, will they infer that trout is also good for you? Or if children believe that too much candy makes them sick, will they conclude the same thing about cake? Given that children may not have information about every single food item they encounter, induction from other foods of the same type could be important to influencing their food choices and eating behavior.

Although the literature on children's naive theory of biology has not directly studied children's use of food categories for induction, several studies have shown children's appreciation for the purpose of eating and its effects. For example, Inagaki and Hatano (2002) found that when 6-year-olds were asked why we eat every day, they tended to favor a vitalistic causal explanation that emphasizes the life force of foods (e.g., they chose an explanation such as "Because our stomach takes in vital power from the food" over an intentional cause ["Because our stomach wants it"] or a mechanistic cause ["Because we take the food into our body after its form is changed in the stomach and bowels"]). Slaughter and Ting (2010) documented similar increases in vitalistic and mechanistic explanations (movement of substances through the body) between the ages of 5 and 8 years. In addition, Wellman and Johnson (1982) found that kindergartners understood the relation between nutritional input and output (e.g., vegetables increase health, strength, and vigor) and the limits of a less varied diet (e.g., eating just one food is not good for you). Raman (2014) found that preschoolers understand that food can affect height and weight (see also Guerin & Thibaut, 2008). Furthermore, Toyama (2000) has shown that 4- to 8-year-olds have an awareness of food digestion and that a lack of food can be damaging to the body. Children's beliefs about food extend to psychological effects. Guerin and Thibaut (2008) and Raman found that children associate foods with emotions such as joy.

Practically speaking, it is essential for planning nutrition education to know what knowledge children bring to the table when learning about foods, especially given the alarming prevalence of overweight and obesity in children in the United States and globally (Centers for Disease Control and Prevention, 2015; World Health Organization, 2015a, 2015b). Data has shown that in France, where the present study was conducted, 15% of children are overweight or obese, compared with 23% of boys and 21% of girls, on average, in Organisation for Economic Co-operation and Development countries (Organisation for Economic Co-operation and Development, 2015).

Unfortunately, a vast number of nutrition education programs are designed mainly to present children with a list of basic facts without embedding them into children's emerging naive theory of biology (for a discussion, see Gripshover & Markman, 2013; Nguyen, McCullough, & Noble, 2011). If children are already using food categories to accumulate knowledge about nutrition by the time they begin school, it is crucial for educators to be aware of this so that they can design programs that capitalize on children's existing abilities and therefore better meet their needs. If children are limited in their use of food categories, then there is an opportunity for programs to intervene. In these ways, schools can play a critical role in building on and supporting young children's existing skills and knowledge to encourage optimal learning of new information about nutrition.

The present studies focused on taxonomic and script categories of food because each category type embodies a different kind of information and has the potential to support a different kind of inference (Ross & Murphy, 1999). Furthermore, there is evidence that children already use both types of category to represent foods (Nguyen & Murphy, 2003). Foods in a taxonomic category (e.g., fruit, vegetables, dairy products) share common features, whereas foods in a script category (e.g., breakfast foods, dinner foods, birthday party foods) play the same role in an event or routine. Past research has documented that children have knowledge of both taxonomic and script organization by ages 2 to 3 years (Nguyen, 2007). That children have knowledge of taxonomic and script organization at an early age suggests that they come to school with concepts of food in hand.

Whether children draw from taxonomic or script categories when reasoning about the effects of eating is an open empirical question that is pertinent to education. Taxonomic proficiency, for example, is often considered a prerequisite for understanding core ideas in biology given that taxonomic categories serve as building blocks for learning more complex biological concepts (Allen, 2015). Script-based categories have practical value (e.g., in knowing what sort of food should be prepared or will be encountered at breakfast) but are probably not a good basis for representing nutritional knowledge. Bacon and cereal are both likely to appear at breakfast, but their nutritional profiles are not similar.

To our knowledge, very few studies have investigated how children use taxonomic and script food categories as a basis for inductive inferences. Nguyen and colleagues have found that by age 4, children use taxonomic food categories to make inferences about the biochemical composition of food but use script food categories to make inferences about the situational contexts in which foods are usually eaten (Nguyen, 2012; Nguyen & Murphy, 2003, Experiment 5). In this research, children were shown picture triads consisting of a target food (e.g., ice cream) along with a taxonomically related match (e.g., milk, another dairy product) and a script-related match (e.g., cake, another birthday party food). When children were told that the target food has novel biochemical stuff inside (e.g., "an ingredient called *pary*") and were asked whether the taxonomically related or script-related match has the same stuff inside, children chose the taxonomic match. In contrast, when children were told that the target food is eaten during a novel "special time" and were asked whether the taxonomically related or script-related match is eaten during the same special time, children chose the script match.

The present research builds on these initial findings to investigate more closely the role of taxonomic food relations in induction about the effects of eating different foods. Taxonomic categories may be particularly relevant to reasoning about nutrition given that foods in these categories share important nutritional features. In this way taxonomic food categories should promote inductive inferences about health outcomes in the human body. Specifically, we examined whether children distinguish taxonomic and script food categories to infer the physiological (hereafter, "biological") and psychological consequences of food consumption. Because foods in a taxonomic category share internal properties, children might expect that these foods give rise to a common impact on the human body. Children may not hold this expectation for script categories because foods in these categories do not have the same internal properties but rather are related by an external event or routine.

A related issue was investigated by Nguyen (2008), namely, whether children use evaluative food categories to reason about nutrition. Foods in an evaluative category receive the same value-laden assessments (e.g., healthy foods, junky foods) and thus can cut across many taxonomic and script categories (e.g., healthy foods may include particular fruits, beverages, and so on). In that study, when children were told that eating a target food caused a novel bodily property (e.g., "Cheetos made Jake's body daxy") and were asked to generalize that property to either an evaluative category match (e.g., ice cream, an unhealthy food) or noncategory match (e.g., fish), children tended to select the evaluative category match. However, because the evaluative food categories tested in that study purposely mixed together foods from several taxonomic and script categories to ensure stimulus diversity (e.g., unhealthy foods included snacks, dairy products, and so on), the contribution of taxonomic versus script category membership to children's inductive inferences about nutrition remains unclear.

Nguyen (2008) did not address the difference between generalization of psychological and biological properties. Many studies have shown that children distinguish biological and psychological properties (Inagaki & Hatano, 1993), but it is unclear how young children would generalize them. For example, they might think that foods elicit only transitory psychological states and so do not prefer taxonomic over script choices. For example, Raman (2014) showed that preschool children did not distinguish between positive and negative long-term psychological outcomes for healthy and unhealthy foods, whereas older children did. It might then be that younger children make no distinction between foods along psychological dimensions (see also Guerin & Thibaut, 2008). Young children might also favor script choices for psychological properties, considering that similar psychological states are elicited by foods that are eaten in the same settings, given that settings themselves may be associated with distinct states.

Our goal was to systematically examine children's use of different categories as a basis of inductive reasoning about biological and psychological effects. To this end we conducted two studies. Experiment 1 included an inductive inference task in which children were told that eating a target food is associated with a particular property and then responded "yes" or "no" to whether a taxonomically related food, script-related food, or unrelated food has the same property. Given that food can have a variety of effects on the body and mind, we tested both biological properties (e.g., stomach ache) and psychological properties (e.g., feelings of happiness). In order to emphasize distinctions among the category types, Experiment 2 used a forced-choice version of the inductive inference task in which children were asked to choose which of two foods (a taxonomically related food and a script-related food) would have the same property as a target food.

In general, if children distinguish the inductive potential of taxonomic and script categories, they should draw from taxonomic categories when reasoning about nutrition. Specifically, for both studies, we predicted that children would make more inductions for taxonomic choices than for the script and unrelated choices for biological properties. It is an open empirical question as to whether psychological properties would show the same effect. Younger children especially might associate psychological effects with different settings and the foods associated with them. Given that knowledge of the ingredients and properties of foods grows over time, we also predicted that taxonomic responding would strengthen with age, at least for biological properties.

Experiment 1: Testing induction in a yes/no paradigm

The purpose of Experiment 1 was to assess children's generalization of a novel property from a target food to other foods that were either taxonomically related, script related, or unrelated to the target food. We tested both biological and psychological properties that have biological underpinnings to capture the myriad effects of foods.

Two hypotheses can be contrasted. The first hypothesis is that children consider food a unified domain, in which all of the members potentially share the same properties. Especially at early ages, children may not strongly differentiate different foods with respect to their properties. If so, they should generalize properties to script-related or even unrelated categories of food. The second hypothesis is that children consider food a complex domain, in which different categories have different kinds of properties. Although the exact way in which children might distinguish categories is not known, one likely possibility is that they will extend both properties to the taxonomically related choices but be more limited in their generalization to script-based categories.

In this experiment, the experimenter presented a target food along with a biological or psychological property. Then the experimenter presented three stimuli, one by one: a taxonomically related food, a script-related food, and an unrelated food. For example, if the target stimulus was strawberry, the taxonomically related food was watermelon, another fruit; the script-related food was whipped cream, something often served with dessert in this population; and the unrelated food was barbecue meat. Children were asked whether each stimulus had the same property as the target stimulus.

Method

Participants

Participants were 11 kindergartners (*M* age = 4.8, range = 3.11–5.4), 17 first-grade children (*M* age = 6.6, range = 5.9–7.2), and 10 third-grade children (*M* age = 9.5, range = 8.9–10.2) . They were from the Burgundy region in eastern France, and French was their native language. They were recruited in schools from middle-class areas. Informed consent was obtained from their parents and from their teachers. Children were tested individually in a quiet room at their school.

Materials

There were 14 stimuli composed of a target food stimulus and a set of three test items, the first taxonomically related to the target, the second script-related to the target, and the third unrelated to the target (see Table 1). Because the stimuli were all foods, in some sense they all belonged to the same category. Taxonomically related items were meant to belong to the same immediate superordinate category (e.g., fruits for strawberry and watermelon), whereas foods connected by a script relation were supposed to appear in the same routine or event (e.g., dessert) but without belonging to the same taxonomic category (e.g., whipped cream is not a fruit). Note that the script-related items were not necessarily familiar pairs, like bread and butter, but were items that often occurred in the same situation. Unrelated foods such as barbecue meat did not belong to the same superordinate category and did not appear in the same event. (Of course, we are not claiming that foods not in the same script are never eaten together. There is nothing to stop someone from eating strawberries with barbecue meat. However, such items were not both in a familiar script event like dessert or birthday parties.)

The stimuli were assessed by 10 psychology students from the University of Burgundy in France. They were given the entire list of foods and were asked to rate to what extent each food belonged to a food category, either a script category (e.g., dessert) or a taxonomic category (e.g., fruits), on a 1-to-7 scale. We selected foods that received a rating of at least 4 out of 7 in the category they were associated with for each category type, either taxonomic or script. (In Experiment 2, the materials had more stringent requirements; see below.) With these ratings, we then constructed a final set of stimuli (see Table 1). Each item was composed of the target food, a taxonomically related food, a script-related food, and an unrelated food. Because the experiment was run in France, the ratings reflect French food choices and associations. For each food, we selected a picture that clearly represented it.

We also constructed a set of seven biological properties and a set of seven psychological properties that a food was said to have on a puppet. The properties were chosen so that they could be understood by young children. The list of properties is presented in Table 2. For half of the participants, the first half of the food trials in Table 1 were associated with biological properties and the second half with psychological properties. For the other half of the participants, the assignment was reversed, thereby counterbalancing food with property type. Thus, each child had seven biological property trials and seven psychological property trials.

Table 1. Categories tested in experiment 1 (translated from the French).

Target	Taxonomic	Script	Unrelated
Strawberry	Watermelon	Whipped cream	Barbecue meat
Chocolate	Pancakes	Lollipop	Peas
Melon	Kiwi	Parma ham	Hamburger
Chicken	Sausage	French fries	Muffin
Milk	Cream	Breakfast biscuit	Steak
Fish (trout)	Breaded fish	Rice	Fruit juice
Soft cheese	Yogurt	Bread loaf	Leek
Ice cream	Yogurt	Chocolate bar	Ham
Broccoli	Tomato	Pork chop	Honey
Croissant	Lemon pie	Chocolate spread	Vegetable sticks
Carrot	Artichoke	Fish (sole)	Brioche
Cake	Grapefruit cake	Milkshake	Plum
Gruyere	Cream cheese	Bread	Cereal bar
Pasta	Rusk	Butter	Biscuit

Table 2. Biological and psychological properties used in Experiments 1 and 2.

Biological properties
- gives Diddl spots on the stomach
- makes Diddl grow her hair faster
- makes Diddl walk very slowly
- helps Diddl breathe
- makes Diddl's face itch
- turns Diddl's toes red
- gives Diddl a tummy ache
Psychological properties
- makes Diddl feel happy
- gives Diddl nightmares, bad dreams
- makes Diddl think sad things
- makes Diddl nervous, stressed
- helps Diddl better remember things she has seen
- makes Diddl smart, clever
- makes Diddl tired, she does not do anything anymore

Procedure

During the task, the experimenter showed a puppet (a Diddl, which is very popular in France) and introduced the task: "We are going to talk about foods." The experiment showed the puppet and said,

> Do you know this puppet, Diddl? Let's talk about where she lives and what she eats. The puppet does not live here. She lives very far away, on a big island, with the sea all around. She lives with other puppets like her. There are many puppets who live there. Doctors have seen that when these puppets eat some types of foods, things might happen in their body (it is also the case for us, when we eat some sorts of foods, we can become sick), or change how they feel, what they think (like us, when we eat some sorts of food, we can feel happy).

Then the experimenter explained,

> Now, I will show you types of foods and I will tell you what this food does to the puppet when she eats it. And then, I will show you other foods and, for each of them, you will tell me whether these foods do the same thing as the first food to the puppet. It is important to know what the foods do to the puppet and to the other puppets who live on the same island, because we want to avoid the puppets from becoming ill.

Then children heard, for example, the following,

> Do you see this food [experimenter introducing the picture of one target food, e.g., a picture of a strawberry]? It's a strawberry. It gives Diddl spots on the stomach and other puppets too. Now, look at this watermelon [e.g., showing the picture of a watermelon], do you think it will give Diddl spots on the stomach too? You are the one who decides. You tell me whether you think the watermelon does the same thing as the first food to the puppet.

These instructions were designed to reassure children that they could respond even if they did not in fact know specifically which foods might result in spots on the stomach, or whatever. Some children are reluctant to make inferences at all, so the "you are the one who decides" instruction was meant to encourage them to give an answer rather than saying, "I don't know." The same question was asked for the two other foods (e.g., the picture of whipped cream and the picture of barbecue meat).

Results

We computed the number of "yes" responses to taxonomic, script, and unrelated choices for the biological and psychological property trials. The maximum score for each category choice per trial type was 7. We ran a 3 × 2 × 3 analysis of variance on children's choices, with age (5-year-olds, 6-year-olds, and 9-year-olds) as a between-subjects factor and type of property (psychological or biological) and type of food choice (taxonomic, script, and unrelated) as within-subjects factors. There was a main effect of type of property, $F(1, 35) = 5.38$, $p < .05$, $\eta_p^2 = .13$, with children answering "yes" more often for the psychological property trials than the biological ones (3.1 vs. 2.7,

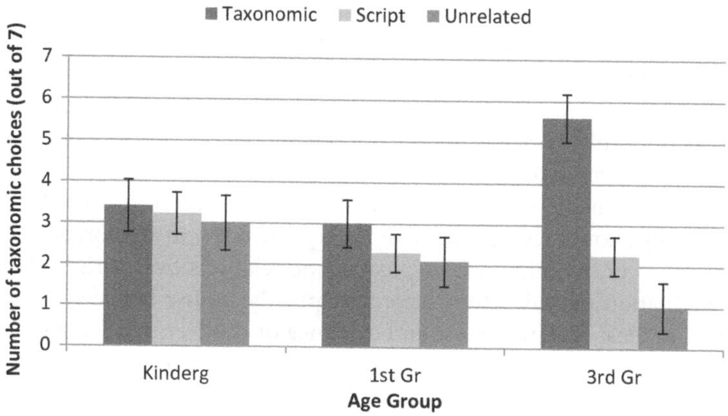

Figure 1. Number of taxonomic choices (out of 7) for each age group for the three types of relations. Error bars show the *SEM*. This interaction shows that third-grade children selected a majority of taxonomically related choices and many fewer script and unrelated choices. The two other age groups did not appear to distinguish the three choices. Kinderg = kindergarten; Gr = grade.

respectively). There was also a main effect of type of food and, more interesting, an interaction between age and type of food: type of food, $F(2, 70) = 31.07$, $p < .001$, $\eta_p^2 = .47$; Age × Type of Food, $F(2, 70) = 12.68$, $p < .001$, $\eta_p^2 = .42$. As shown in Figure 1, 9-year-old children answered "yes" to taxonomic choices more often than the other two age groups and had the smallest number of "yes" responses to unrelated choices (Tukey's honestly significant difference [HSD], $p < .01$). The two other age groups did not discriminate between the three choices (Tukey's HSD, *ps* > .05). The Type of Property × Type of Food interaction and the triple interaction were not significant: Type of Property × Type of Food, $F(2, 70) = 0.41$, $p > .5$; triple interaction, $F(4, 70) = 1.5$, $p > .10$. The sample size was a bit small in the youngest group, but its means were right in the middle of the scale and did not look at all like the results of the oldest group.

Discussion

The most noteworthy result was the interaction between age and type of food, showing that only 9-year-old children generalized properties to taxonomically related categories more often than to the two other category types (script and unrelated categories). Note that 9-year-olds rarely considered unrelated categories as valid choices. This pattern is consistent with the second hypothesis described earlier, that children distinguish different food categories. It is interesting that these children did not make different choices for psychological and biological properties—both were generalized based on taxonomic categories, perhaps suggesting that 9-year-olds see a biological basis for both kinds of food effects.

By contrast, the two younger age groups selected the three types of food at the same rate. Kindergartners selected the three options essentially at chance (3.5 "yes" out of 7). One interpretation of this result is that young children treated all of the food categories as equivalent. This does not seem consistent with the existing literature. For example, Nguyen (2008) found that children treated the evaluative categories of healthy foods as distinct from junky foods, selectively attributing different properties to each type of food. However, evaluative categories are defined around one property: health. The present experiment tested the more general case of script-based categories, which children apparently are less likely to use as a basis for induction, at least for the types of properties we tested. Nguyen and Murphy (2003) found that children generalize situational features to other members of script-based categories. Our data show that neither biological nor psychological properties are strongly generalized on the basis of such categories.

It might be that the yes/no procedure underestimates children's competence. Perhaps it is difficult for them to judge whether a novel category has the same property as the target category because all of the items are foods, and it is difficult to decide at which level of categorization the target property should be generalized (e.g., from strawberries to other berries? To all fruit? To all plant-based foods?). One way to test children's sensitivity to levels of categorization would be to use a forced-choice task, in which they have to choose one food out of two, the first one being taxonomically related and the second one being script related. If the younger children truly do not distinguish the different categories, they should choose them about equally often in the forced-choice task, for both kinds of property. However, if children favor taxonomic relations over script relations as a basis for their inductions, a situation in which they have to explicitly choose between the two options might reveal their preferences. Past research has found evidence of children's inductive selectivity using this forced-choice technique (see Nguyen & Murphy, 2003). Therefore, we used this paradigm in Experiment 2. We also improved the materials in the sense that we more strictly distinguished taxonomically and script-related options.

Experiment 2: Forced choice

Experiment 1 revealed that younger participants did not generalize properties to the taxonomic choice more often than the other two choices. This result suggests that 6-year-olds may consider the different types of classification possibilities as equivalent: Belonging to the same taxonomic or the same script category is sufficient to generalize a food's properties, both psychological and biological. However, in fact, younger children did not generalize the target property strongly to any categories, as shown by the mean percentage of choice, which remained below 3.5 out of 7. One interpretation is that younger children were conservative in positively claiming that a food would have a property that they had never heard of. This could have made it difficult to find differences among the category types.

Another way to study whether children can distinguish between taxonomic and script choices is to provide more constraints on the task. Thus, in Experiment 2, children chose between a taxonomic and a script choice in each induction question. This avoids the conservatism problem, because children are not choosing between "yes" and "no" but between the two foods. We can contrast three hypotheses. First, if children believe that properties should be shared by foods that are closely related in a taxonomy, they will generalize the novel property to taxonomic choices. Second, if they think that foods that are eaten together should share properties, then they will choose the script choice, possibly more for the psychological properties than for the biological properties. Third, if they do not really distinguish between foods, they will generalize the property roughly equally to the taxonomic and script choices. In this experiment, we did not include 8- to 9-year-olds, who already gave a majority of taxonomic choices in Experiment 1. Instead, we also included 4-year-olds because we did not know a priori at what age children would fail to discriminate between taxonomic and script choices.

Method

Participants

French-speaking children were recruited at their school from the same population as Experiment 1. There were 20 four-year-olds (*M* age = 4;9, range = 4;4–5;4), 19 five-year-olds (*M* age = 5;9, range = 5;4–6;0), and 25 six-year-olds (*M* age = 6;4, range = 6;0–7;0).

Materials

We modified and improved the set of foods used in Experiment 1 (see Table 3). Again, university students saw pairs of food and had to rate whether they belonged to the same taxonomic categories on a 1-to-7 scale. Students were told that the two foods had to belong to the same immediate superordinate category (e.g., apple and pear belong to the fruit category). In a similar way, all of the pairs were rated in terms of script relationships. Participants were told that the food had to be used

Table 3. List of Items Used in Experiment 2.

Target	Taxonomic	Script
Strawberry	Coconut	Whipped cream
Cutlet	Roulade	Pasta
Melon	Avocado	Parma ham
Chicken	Sausage	French fries
Glass of milk	Yogurt	Breakfast biscuits
Fish (trout)	Breaded fish	Rice
Soft cheese	Hard cheese	Bread
Ham	Steak	Green beans
Broccoli	Pumpkin	Pork chop
Croissant	Lemon tart	Hot chocolate
Carrot	Asparagus	Fish (sole)
Cake	Cake with prunes	Compote
Gruyere	Yogurt	Bread
Pasta	Rice	Butter

in the same routine or event (e.g., strawberry and whipped cream belong to the dessert category and are served together, at least in France).

For the taxonomically related list, we kept pairs of foods that received high ratings on the taxonomic relation and low ratings on the script relation. There were 14 such pairs. Apart from one pair of stimuli, all of the differences between the two scales were beyond 3.3, with a mean difference of 3.8 (i.e., taxonomic ratings were on average 3.8 higher than script ratings for the same pair). The one exception had a difference of 2.8. The ratings for the script-related items were on average 3.7 points higher than the taxonomic rating for the same pair. Apart from one item (difference = 2.2), all of the differences between the two scales were greater than 3. We again selected pictures that clearly illustrated each item.

The biological and psychological properties were the same as in Experiment 1.

Procedure

We used the same procedure as in Experiment 1, except that the yes/no procedure was replaced by a forced-choice procedure in which participants were shown a picture of a food that was associated with a biological or a psychological property. Then they were asked which one of two foods (taxonomic or script) would produce the same effect as the target food. For example, the experimenter told children, "See this food. It is a strawberry [the experimenter shows a picture of a strawberry]. It makes Diddl smart. See these other foods here [the experimenter shows a picture of each of these two foods simultaneously], which one do you think will also make Diddl smart?"

Results

We computed the number of taxonomic choices in the forced-choice task. As in Experiment 1, the maximum score for each trial type was 7. We ran a 3 × 2 analysis of variance on children's choices, with age (4-, 5-, and 6-year-olds) as a between-subjects factor and type of property (psychological or biological) as a within-subjects factor on the number of taxonomic choices. There was no effect of type of property (biological vs. psychological). However, there was a main effect of age, $F(2, 52) = 23.7$, $p < .0001$, $\eta_p^2 = .22$. Mean taxonomic choices were 5.5 for 6-year-olds, 4.4 for 5-year-olds, and 4.1 for 4-year-olds. Post hoc tests (Tukey's HSD) showed that the 6-year-olds differed from the two other groups, which did not differ significantly one from the other. There was no interaction between type of property and age, $F(2, 52) < 1$.

For the 6-year-olds, taxonomic selection for both psychological ($M = 5.3$) and biological ($M = 5.6$) properties differed from chance (3.5) by t test ($ps < .001$). That is, they chose the taxonomically related food more often than the script-related food. For the 5-year-olds, taxonomic selection for both psychological and biological properties differed from chance ($ps < .05$, with 4.4 for both biological and psychological properties), whereas for 4-year-olds, psychological properties differed statistically from chance (*ps* < .05, *M* = 4.3) but biological properties did not (*p* > .05, *M* = 3.9).

Discussion

In contrast to Experiment 1, this forced-choice experiment showed that 5- and 6-year-old children selected the taxonomic choice over the script choice. Four-year-olds were at chance for biological properties, whereas psychological properties were beyond chance. Although the proportion of taxonomic responses was not very different for the two property types (4.3 vs. 3.9 for psychological vs. biological), it is somewhat surprising that biological properties did not give the stronger taxonomic preference. These results are consistent with previous findings showing, with different tasks, that children know that food can have both biological and psychological effects (Guerin & Thibaut, 2008; Raman, 2014). They are also consistent with the finding in Experiment 1 that younger children do not clearly distinguish the two kinds of properties in their inductions, even with a more sensitive forced-choice task. One might have thought that kindergartners simply did not understand the induction task in Experiment 1. However, the same-aged children did respond selectively to category members when given the questions in a forced-choice format, suggesting that they do understand about induction of properties. More generally, the induction task has been successfully used with children younger than the ones in our study (see Murphy, 2002).

General discussion

In two experiments, we investigated children's induction of biological and psychological properties toward taxonomically and script-related categories. The main question was whether young children would distinguish these two types of categories, and at what age. Another question was whether the two types of properties would be generalized in the same way.

Experiment 1, a yes/no task, showed that the two younger groups generalized properties equally often to taxonomic, script, and unrelated choices. Older children favored taxonomic categories. Using a different test, Experiment 2 showed that even kindergartners preferred taxonomic over script choices. This was true for both psychological and biological properties, although 4-year-olds chose taxonomically related foods above chance for psychological properties only. In sum, children reliably chose taxonomically related foods for induction, starting at age 4 years for the psychological properties and at 5 years for the biological properties.

How can we account for the differences between the results of the two experiments? In Experiment 1, each food was assessed independently, one by one. Thus, participants were not required to compare the foods against each other to find one that would be more plausible. By definition all of the choices were foods and thus taxonomically related to the target food at some level, so perhaps children did not think very carefully about their particular relation to the target. Furthermore, it is possible that children were reluctant to positively assert that a food had an unfamiliar property, thereby lowering their overall induction rate. In Experiment 2, children always chose one food over the other, and so this conservatism did not operate. In this context, children preferred taxonomic choices over script choices. Finally, the items were more strictly selected in Experiment 2, and the stronger taxonomic relations may have exerted a stronger effect than in Experiment 1.

A priori, one could have hypothesized that children would judge foods belonging to the same meal or dish (whipped cream with strawberries) to share the same properties, especially for psychological properties. Although it seems clear from previous studies that children of these ages do have script-based categories (Nguyen & Murphy, 2003), they apparently are not a strong basis for induction of the effects of eating food. This may be helpful from a practical standpoint, as nutrition instruction may not have to overcome biases such as children thinking that all breakfast foods have protein or all birthday foods are bad for you.

How do these results relate to those of previous studies? The present research shows that even preschoolers can successfully extend a property to specific, taxonomically related categories. Guerin and Thibaut (2008) and Raman (2014) found a different pattern of results, but with different tasks. Guerin and Thibaut showed that psychological properties were less distinctively associated with specific categories of foods. For example, they asked kindergartners and first-grade children whether increasing the consumption of different types of food (fruits, vegetables, dairy, meats) would change puppets' psychological and biological properties. Results revealed no difference between foods for psychological properties, although children mentioned that fruits could affect biological properties (health). Raman found that preschool children attributed the same number of positive and negative psychological properties to healthy and unhealthy foods. Possibly the present task was easier than those tasks. In those studies, participants had to assess whether a type of food could influence puppets. This required referring to explicit biological knowledge about the nature or the mechanism underlying this influence. In Experiment 2, by contrast, they had to decide to which type of category they should extend a property but not how this property works.

Previous research has also shown that 4-year-olds can sort foods into healthy and junky foods (Nguyen, 2007). In an induction task similar to ours, Nguyen (2008) showed that 4-year-olds generalized a fictitious property (e.g., "makes Jim's body daxy") given to a healthy food such as an apple to another healthy food rather than to a junk food. The main difference with our research is that Nguyen's foods were restricted to a conflict between two specific categories, healthy versus junk food, and to fictitious properties. Because the properties had no reference, the question of their type (psychological or biological) was moot. By contrast, we contrasted two general types of food categories (script or taxonomic) that participants could not reduce to any specific property (such as healthy vs. unhealthy), and the properties were unfamiliar but understandable reactions to eating the food. We also found that both psychological and biological properties were generalized in the same way.

Nguyen and Murphy (2003) also used a forced-choice inductive inference task in which participants were asked to choose whether a script or a taxonomically related choice had the same property as a target food (Experiment 5). They found that 4-year-olds chose the taxonomically related food less than chance for both biochemical properties (e.g., "bread has pary in it") and situational properties (e.g., "cake is eaten on a special holiday called *dax*"), whereas 7-year-olds and adults were selective in that they made taxonomic choices for biochemical properties and script choices for situational properties. Our study examined whether children would similarly infer biological and psychological reactions to eating taxonomically or script-related foods. The results showed that both types of property were extended to taxonomically related foods. In the Nguyen and Murphy study, situational properties were specifically matched to script-based categories in that both referred to contexts in which foods are encountered. Children may well be correct in thinking that if one kind of meat causes a psychological reaction, then another kind of meat might also. If eating cereal in the morning makes you depressed, it does not necessarily follow that eating bacon will have the same reaction. Overall, it seems that children more easily associate taxonomic categories to differences in food that are intrinsic, whether biological or psychological, consistent with theorizing about taxonomic categories more generally (Ross & Murphy, 1999).

Finally, even though a comparison of the two experiments was not an a priori goal of this research, it is interesting to note that children's choices were more evenly spread among the available options (including unrelated choices) in the first experiment than in the second experiment. This might mean that when participants have to choose between two foods, the comparison between the two leads to deeper encoding, which favors taxonomic choices. The positive effect of comparison has been widely demonstrated and has been shown to favor taxonomic choices over other encodings (perceptual or thematic; Augier & Thibaut, 2013; Gentner & Namy, 1999; Thibaut & Witt, 2015). Comparisons are useful in that the alignable differences between the items being compared are highlighted and more readily identified (Markman & Gentner, 1993). In this respect, the yes/no paradigm in which each food is evaluated by itself is less constraining and so might not lead children to think of the relevant properties.

This methodological difference has educational consequences. It suggests that when one wants children to learn properties of types of foods, a comparison format might be more efficient than a no-comparison format. Asking children to compare a dessert to a salad may cause them to attend to the salient differences between them, which would not at all be considered if they simply thought about one alone. Our second experiment shows that children between 4 and 5 are able to understand that related foods might share the same consequences. At about the same age, children using the same forced-choice format are able to treat foods that are defined by a salient property (healthy) as an homogeneous category (Nguyen, 2008). As shown by all of these studies, formats that explicitly encourage comparisons between foods might promote the emergence of conceptually salient dimensions.

In addition to highlighting differences, comparison can help children learn commonalities of category members. One way to enhance taxonomically based generalizations in children would be to show pairs or triplets of taxonomically related foods (as in Namy & Gentner, 2002, or Thibaut & Witt, 2015) about which one wishes to teach a property (chemical, biological, health, etc.), such as "These ones are poisonous" or "These have sugar inside" or "These ones give people headaches." The comparison story suggests that once the subset is attributed a target property, young children will compare the items and will be more efficient in generalizing the target property to taxonomically related items than in the case of a single item (Namy & Gentner, 2002). As pointed out by Thibaut and Witt (2015), the optimal number of items involved in the comparison is an empirical question. In a similar way, the role of contrasting items (e.g., "This one is daxy but this one is not daxy") is also an empirical question. Augier and Thibaut (2013) showed that contrasting items could disrupt younger children's learning and generalization, whereas it could contribute to better generalization for older children. As a general principle, comparison can be helpful, but one must take care in how it is implemented.

Finally, both Experiments 1 and 2 suggest that children have a general preference for using taxonomic categories in induction, as adults often do (see Ross & Murphy, 1999). This is probably a good thing for educational purposes, given that most benefits and problems associated with foods are based on their chemical composition, which in turn is most related to their taxonomic categories. For example, our children did not seem tempted to say that if strawberries had a given effect, that whipped cream, often consumed with the strawberries, would have the same effect. If confirmed, this finding suggests that nutritional education has a natural basis in children's category-based induction even before the elementary school years.

ORCID

Jean-Pierre Thibaut http://orcid.org/0000-0002-3645-6742

Acknowledgments

We wish to thank Hélène Guérin and Pauline Bonnet for their help in stimulus construction, data gathering, and analysis.

References

Allen, M. (2015). Preschool children's taxonomic knowledge of animal species. *Journal of Research in Science Teaching, 51*, 107–134. doi:10.1002/tea.21191

Augier, L., & Thibaut, J. P. (2013). The benefits and costs of comparisons in a novel object categorization task: Interactions with development. *Psychonomic Bulletin & Review, 20*, 1126–1132. doi:10.3758/s13423-013-0436-x

Centers for Disease Control and Prevention. (2015, August 5). *Childhood overweight*. Retrieved from http://www.cdc.gov/obesity/childhood/

Gelman, S. A. (1988). The development of induction within natural kind and artifact categories. *Cognitive Psychology, 20*, 65–95. doi:10.1016/0010-0285(88)90025-4

Gelman, S. A., & Davidson, N. S. (2013). Conceptual influences on category-based induction. *Cognitive Psychology, 66*, 327–353. doi:10.1016/j.cogpsych.2013.02.001

Gelman, S. A., & Kalish, C. W. (2006). Conceptual development. In D. Kuhn & R. Siegler (Eds.), *Handbook of child psychology: Vol. 2. Cognition, perception and language* (pp. 687–733). New York, NY: Wiley.

Gentner, D., & Namy, L. L. (1999). Comparison in the development of categories. *Cognitive Development, 14*, 487–513. doi:10.1016/S0885-2014(99)00016-7

Girgis, L., & Nguyen, S. P. (2015, March). *Children's judgments regarding the causal mechanisms underlying food origins.* Poster presented at the biennial meeting of the Society for Research in Child Development, Philadelphia, PA.

Gripshover, S. J., & Markman, E. M. (2013). Teaching young children a theory of nutrition: Conceptual change and the potential for increased vegetable consumption. *Psychological Science, 24*, 1541–1553. doi:10.1177/0956797612474827

Guerin, H., & Thibaut, J. P. (2008). The development of food representations in children 4 to 12 years. *Enfance, 60*, 251–260. doi:10.3917/enf.603.0251

Hayes, B. K. (2007). The development of inductive reasoning. In A. Feeney & E. Heit (Eds.), *Inductive reasoning: Experimental, developmental, and computational approaches* (pp. 25–54). New York, NY: Cambridge University Press.

Inagaki, K., & Hatano, G. (1993). Young children's understanding of the mind-body distinction. *Child development 64*, 1534–1549.

Inagaki, K., & Hatano, G. (2002). *Young children's intuitive thinking about the biological world.* New York, NY: Psychology Press.

Inagaki, K., & Hatano, G. (2006). Young children's conception of the biological world. *Current Directions in Psychological Science, 15*, 177–181. doi:10.1111/j.1467-8721.2006.00431.x

Markman, A. B., & Gentner, D. (1993). Structural alignment during similarity comparisons. *Cognitive psychology, 25* (4), 431–467.

Murphy, G. L. (2002). *The big book of concepts.* Cambridge, MA: MIT Press.

Namy, L. L., & Gentner, D. (2002). Making a silk purse out of two sow's ears: Young children's use of comparison in category learning. *Journal of Experimental Psychology: General, 131*, 5–15. doi:10.1037/0096-3445.131.1.5

Nguyen, S. P. (2007). Cross-classification and category representation in children's concepts. *Developmental Psychology, 43*, 719–731. doi:10.1037/0012-1649.43.3.719

Nguyen, S. P. (2008). Children's evaluative categories and inductive inferences within the domain of food. *Infant and Child Development, 17*, 285–299. doi:10.1002/icd.553

Nguyen, S. P. (2012). Inductive selectivity in children's cross-classified concepts. *Child Development, 83*, 1748–1761. doi:10.1111/j.1467-8624.2012.01812.x

Nguyen, S. P., McCullough, M. B., & Noble, A. (2011). A theory-based approach to teaching children about health: A recipe for understanding. *Journal of Educational Psychology, 103*, 594–606. doi:10.1037/a0023392

Nguyen, S. P., & Murphy, G. L. (2003). An apple is more than just a fruit: Cross-classification in children's concepts. *Child Development, 74*, 1783–1806. doi:10.1046/j.1467-8624.2003.00638.x

Organisation for Economic Co-operation and Development (2015, October 10). *Obesity and the economics of prevention: Fit not fat—France key facts.* Retrieved from http://www.oecd.org/els/health-systems/obesityandtheeconomicsofpreventionfitnotfat-francekeyfacts.htm

Raman, L. (2014). Children's and adults' understanding of the impact of nutrition on biological and psychological processes. *British Journal of Developmental Psychology, 32*, 78–93. doi:10.1111/bjdp.12024

Ross, B. H., & Murphy, G. L. (1999). Food for thought: Cross-classification and category organization in a complex real-world domain. *Cognitive Psychology, 38*, 495–553. doi:10.1006/cogp.1998.0712

Slaughter, V., & Ting, C. (2010). Development of ideas about food and nutrition from preschool to university. *Appetite, 55*(3), 556–564. doi:10.1016/j.appet.2010.09.004

Thibaut, J. P., & Witt, A. (2015). Young children's learning of relational categories: Multiple comparisons and their cognitive constraints. *Frontiers in Psychology, 6*, 643. doi:10.3389/fpsyg.2015.00643

Toyama, N. (2000). What are food and air like inside our bodies? Children's thinking about digestion and respiration. *International Journal of Behavioral Development, 24*, 222–230. doi:10.1080/016502500383359

Wellman, H. M., & Gelman, S. A. (1997). Knowledge acquisition. In D. Kuhn & R. S. Siegler (Eds.), *Handbook of child psychology: Vol. 2. Cognition, perception, and language* (6th ed., pp. 523–573). Hoboken, NJ: Wiley.

Wellman, H. M., & Johnson, C. N. (1982). Children's understanding of food and its functions: A preliminary study of the development of concepts of nutrition. *Journal of Applied Developmental Psychology, 3*, 135–148. doi:10.1016/0193-3973(82)90024-7

World Health Organization (2015a, October 6). *The challenge of obesity in the WHO European region and the strategies for responses.* Retrieved from http://www.euro.who.int/__data/assets/pdf_file/0010/74746/E90711.pdf

World Health Organization (2015b, October 6). *Childhood overweight and obesity.* Retrieved from http://www.who.int/dietphysicalactivity/childhood/en/

Index

For Product Safety Concerns and Information please contact our
EU representative GPSR@taylorandfrancis.com Taylor & Francis
Verlag GmbH, Kaufingerstraße 24, 80331 München, Germany